PHYSICS OF THE
UNIVERSE

PHYSICS OF THE UNIVERSE

Mendel Sachs

University at Buffalo, The State University of New York, USA

Imperial College Press

Published by

Imperial College Press
57 Shelton Street
Covent Garden
London WC2H 9HE

Distributed by

World Scientific Publishing Co. Pte. Ltd.

5 Toh Tuck Link, Singapore 596224

USA office: 27 Warren Street, Suite 401-402, Hackensack, NJ 07601

UK office: 57 Shelton Street, Covent Garden, London WC2H 9HE

Library of Congress Cataloging-in-Publication Data
Sachs, Mendel.
   Physics of the universe / Mendel Sachs. -- 1st ed.
     p. cm.
   Includes bibliographical references and index.
   ISBN-13: 978-1-84816-532-8 (hardcover)
   ISBN-10: 1-84816-532-3 (hardcover)
   ISBN-13: 978-1-84816-604-2 (pbk)
   ISBN-10: 1-84816-604-4 (pbk)
   1. Cosmology. 2. Unified field theories. 3. General relativity (Physics) I. Title.
   QB981.S253 2010
   523.1--dc22

                        2009048635

British Library Cataloguing-in-Publication Data
A catalogue record for this book is available from the British Library.

Typeset by Stallion Press
Email: enquiries@stallionpress.com

Printed in Singapore.

Dedicated to

Albert Einstein
Physicist — Philosopher — Humanitarian

And his Quest for Fundamental Truths of the Universe

Contents

Preface   xi

Acknowledgements   xvii

Chapter 1. Physics of the Universe   1

Introduction   1

Is Newton's Theory an Explanation of Gravity?   4

The Expanding Universe   5

The Oscillating Universe Cosmology   6

The Theory of General Relativity   7

The Role of Space and Time   8

Geometry and Matter   11

Generalization of Einstein's Field Equations   13

A Unified Field Theory   16

Chapter 2. A Language of Cosmology:
The Mathematical Basis of General Relativity   18

Introduction   18

Einstein's Tensor Formulation   19

The Riemann Curvature Tensor   19

The Geodesic Equation   21

The Vacuum Equation 21

   The Schwarzschild Solution 22

   The black hole 23

The Crucial Tests of General Relativity 24

The Logic of the Spacetime Language 25

Chapter 3. A Unified Field Theory in General Relativity:
Extension from the Tensor to the Quaternion
Language 27

Introduction 27

Factorization of Einstein's Tensor Field Equations 28

The Riemann Curvature Tensor in Quaternion Form 29

   Spin-affine connection 30

   Spin curvature 31

The Quaternion Metrical Field Equations 32

A Symmetric Tensor-Antisymmetric Tensor
Representation of General Relativity — Gravity and
Electromagnetism 32

The Einstein Field Equations from the Symmetric
Tensor Part 33

The Maxwell Field Equations from the Antisymmetric
Tensor Part 33

Conclusions 35

Chapter 4. An Oscillating, Spiral Universe Cosmology 37

Introduction 37

   The oscillating universe cosmology 39

   Equations of motion in general relativity 39

Dynamics of the Expansion and Contraction of the Universe   41

    The geodesic equation in quaternion form   41

Dynamics of the Oscillating Universe Cosmology   44

Derivation of the Hubble Law as an Approximation   46

The Spiral Structure of the Universe   46

Concluding Remarks   49

Chapter 5. Dark Matter   **51**

Introduction   51

The Field Equations and the Ground State Solution
for the Bound Particle-Antiparticle Pair   53

Separation of Matter and Antimatter in the Universe   56

Olber's Paradox   57

Chapter 6. Concluding Remarks   **60**

Black Holes   60

Pulsars   61

On the Human Race and Cosmology   62

Chapter 7. Philosophical Considerations   **64**

On Truth   64

Positivism versus Realism, Subjectivity versus Objectivity   67

On Mach's Influence in Physics and Cosmology   70

    The quantum mechanical limit   70

    The Mach principle   71

    The Mach principle and a unified field theory   72

References and Notes   74

Postscript   **77**

Physics in the 21st Century   78

Holism   95

The Universe   99

The Mach Principle and the Origin of Inertia from General
Relativity   108

Index   125

Preface

This monograph presents a fresh look at the problems of the physics of the universe — the subject of cosmology — compared with other contemporary cosmological theories. Its main purpose is to fully exploit the most fundamental expression of Einstein's theory of general relativity as the mathematical description (language) and the explanation of the physical behavior of the material universe. Contemporary quantum theory, the 'standard model' of elementary particle physics and thermodynamics, play no fundamental role in this analysis.

Physics is an empirical science. That is, its claims to understand the material behavior of the objects of the universe, with a logically consistent theory, must be backed by agreement between the theoretical predictions and the empirical facts. Newton's theory of universal gravitation was believed to be true for the three centuries that preceded Einstein's discovery of general relativity. But in the context of cosmology, Newton had no empirical backing for anything beyond our solar system — an infinitesimal portion of the universe! In the 20th century new experimental facts came to light on the problem of cosmology that did not fit Newton's theory. Primary was Edwin Hubble's discovery of the expansion of the universe. More recently, high-resolution instrumentation for observing the universe, such as the Hubble telescope, revealed observations unknown to Newton or his contemporaries, and his followers during the next three centuries. It has led to the revelation of a universe composed of an infinitude of galaxies, in addition to our own galaxy, the Milky Way, the spiral structures and rotations of most of the galaxies, the clustering of the galaxies, the background radiation field of the universe, and to exotic stars, such as the quasars and pulsars.

In the early decades of the 20th century, Einstein's theory of general relativity reproduced all the successful results of Newton's theory of universal gravitation, as well as new empirical predictions of gravitational effects not at all predicted by the classical theory. Thus Einstein's general relativity superseded Newton's theory as an explanation of the phenomenon of gravity. Einstein's theory of gravitation was entirely different to that of Newton, conceptually and mathematically. That is to say, Newton's theory of universal gravitation *is not included* in Einstein's theory of general relativity. Rather, Newton's formalism is not more than a mathematical approximation for that of Einstein's theory of general relativity.

Newton's theory is based on a model of the universe as an open system, on atomism and on the concept of action-at-a-distance. Newton himself was dissatisfied with this as an *explanation* of gravitation, but he accepted it as a useful *description*, for his day. In his theory of gravitation, space and time are separate measures in the description of the gravitational force. In contrast, Einstein's theory entails the fusion of the space and time measures in the language of the laws of nature. In this generalization of the language, the space and time measures cannot be objectively separated. They form a fused spacetime, wherein a spatial (or temporal) measure in the language of a law of nature in one frame of reference is a mixture of a spatial and a temporal measure in a different reference frame where the same law is to be compared. Here is the idea of the universe as a closed system, where forces between matter components propagate at a finite speed. These matter components of the universe, in turn, are not separate, singular entities. Rather, they are the distinguishable correlated modes of the closed, continuous matter of the universe. Instead of atomism, this view is then based on the continuous, holistic field concept. (It is compatible with Mach's interpretation of the inertial mass of matter, named by Einstein, 'the Mach principle'.) *This is a genuine paradigm change from the atomistic view that has dominated physics for the many centuries since ancient Greece. An exception in the history of science was Michael Faraday's support of the continuous field concept, in the 19th century.*

One of the important predictions of the Einstein theory was that in the description of the laws of nature, because the space and time measures are not separate, but are rather a fused spacetime measure, there is the implication that the objects of the universe are not in stationary orbits about other objects, as Newton's theory would hold. Instead, Einstein's theory is in agreement with the Hubble observation that the matter components of the universe are not in stationary orbits, rather they are moving away from all other material components of the universe, i.e. that *the universe is expanding.* Or in a different phase of an oscillating universe cosmology, they are moving toward all other matter components in a contraction phase of the universe.

Are there any deficiencies in Einstein's theory of general relativity, based on its own premises? Yes, there are. It was Einstein who said that to fully exploit the theory of general relativity, not only must one study its geometrical basis, but also its algebraic basis. The latter refers to the most general expression of the symmetry group that underlies this theory of matter. In my studies I have found that this is a *continuous group (indeed, it is a continuous group of analytic transformations — thus a 'Lie Group'. I have called it the 'Einstein Group').* But the group of transformations that leave Einstein's tensor field equations covariant (i.e. frame-independent) are not only continuous, as they must be, but they are also reflection symmetric, which is not required. Thus Einstein's tensor equations are *not* the most general form for his theory of general relativity.

What I have shown in this monograph is that when the reflection symmetry elements of the underlying symmetry group are removed, Einstein's (reducible) tensor field equations factorize to an (irreducible) quaternion field equation. The 10-component, symmetric metric tensor $g^{\mu\nu}(x)$ of Einstein's original formulation of general relativity is then replaced with the 16-component quaternion metric field $q^{\mu}(x)$. This is covariant as a four-vector under the continuous transformations that characterize the Einstein group, but it is not covariant under the discrete reflection elements that characterize the symmetry of Einstein's tensor field equations. The quaternion field $q^{\mu}(x)$ is a four-vector in which each of its four components is quaternion-valued. Thus this field has $4 \times 4 = 16$ independent components. The new quaternion metrical field equations are then 16 in number. I have shown in Chapter 3 (pp. 27–36) that this leads to a unified field theory — by iteration it is seen that ten of the components correspond exactly with Einstein's tensor formulation, thus explaining gravity. The remaining six equations give the Maxwell field equations in the three electric and three magnetic field variables, thus explaining electromagnetism. **This is the unified field theory that was sought by Einstein and Schrödinger.**

In Chapter 4 (pp. 37–50) it is shown that under the appropriate conditions, the new quaternion formulation of general relativity leads to: 1) The (empirically correct) Hubble law, as a non-covariant approximation for the dynamics of an expanding universe and 2) the spiral structure of an oscillating universe, alternating between expansion and contraction, in the never-ending cycles of the universe. Thus, with this view, there is no absolute (singular) 'beginning' of the universe. An absolute temporal beginning of the universe is indeed ruled out as incompatible with the relativity of the time measure in Einstein's theory of relativity. Further, the spacetime is necessarily curved, *everywhere*. The flat spacetime is only an ideal, unreachable limit that simply characterizes the universe as a vacuum *everywhere*.

In Chapter 5 (pp. 51–59) it is shown that a viable candidate for the dark matter of the universe is a dense sea of particle-antiparticle pairs (electron-positron and proton-antiproton) in a particular (*derived*) bound state of null energy, momentum and angular momentum — it is the true ground state of the pair; the sea of such pairs in this state is likened in other theories to the 'vacuum state'.

Many of these results are derived in more detail in previous publications of this author. The main purpose of this presentation is to focus discussion primarily on the problems of cosmology and to indicate their resolutions in a maximally explanatory way.

There are some brief, concluding remarks in the Chapters 6 and 7 on the subject of black holes and pulsars, as well as philosophical considerations.

Summing up, the language of the physics of the universe — the language of cosmology — is presented here in terms of Einstein's theory of general relativity, expressed in its *irreducible form* by removing the reflection symmetry elements from the underlying group (thus yielding the irreducible symmetry group — the 'Einstein group'). This leads to a unified field theory in terms of the generalization of the symmetric tensor representation (10 independent field equations) to the quaternion representation (16 independent field equations). This unification explains, in fundamental terms, gravitation and electromagnetism in a single formalism.

The (irreducible) quaternion formalism, as the language for cosmology, then leads to a new view of the universe. The dynamics of the universe is seen in terms of a cyclical, oscillating model, alternating between expansion and contraction. Further, this model is not in terms of an isotropic and homogeneous matter distribution of the universe, as the present-day cosmological theories contend. Rather, it is in terms of a spiral, rotating material universe, in a curved spacetime. The Hubble law is predicted here as an approximation for the dynamics of any expansion phase of the oscillating universe. With the view of 'dark matter' developed here, a prediction of this model is the empirically observed separation between matter and antimatter in the universe, occurring during the (non-singular!) 'big bang' that initiates any expansion phase. In contrast with present-day models that evoke concepts of particle physics — the quantum theory, the standard model, the string theories — these play no role in this dynamical representation of the physics of the universe. It is explained here strictly in terms of the theory of general relativity and its irreducible expression, based on the underlying *principle of covariance*.

Mendel Sachs
Buffalo, New York
July, 2009

Acknowledgements

My research program on cosmology, based on general relativity theory and unified field theory, was inspired by the writings of Albert Einstein and Erwin Schrödinger. I gained also from the 19th-century writings of William Rowan Hamilton and his discovery of the quaternion algebra. I had the good fortune to spend some time in discussion on these subjects with Cornelius Lanczos, at the Dublin Institute for Advanced Studies. Last, but not least, I thank my son, Robert, for persuading me to write this book.

1

Physics of the Universe

Introduction

Physics is the science of inanimate matter. Cosmology is the part of this science that deals with the universe as a whole. It is the oldest and the youngest branch of physics. It is the oldest because the heavens were studied in ancient times, in Greece and in Asia, and other parts of the world. It is the youngest because it has been re-invigorated in recent times due to observations with new, high-resolution astronomical instrumentation (such as the Hubble telescope) and theoretical analyses in the context of current thinking in particle physics and relativistic dynamics. Voluminous works have been written on the order of the night sky. (*The Greek word*, *'Cosmology', means 'order' (logos) of the cosmos*.)[1]

Astronomical laboratories have been constructed since the ancient times to study this order. Examples include the Stonehenge monument, built by the ancient Britons thousands of years ago, and similar ancient astronomical viewing sites in India, China, Australia, Peru, Mexico and from other cultures in the different corners of the world, designed by the ancient and aboriginal peoples to see the star formations and their locations, the locations of the sun and the moon, at the different times of the year. In these ancient viewings, there was no magnification.

Galileo, in the 16th century, was the first astronomer to use magnification, utilizing the telescope — a series of lenses that he contrived to view the heavens.[2] Focusing mainly on our solar system, he saw the moons of Jupiter, the sun spots, the landscape of the moon, and he verified the conclusion of Copernicus that *the earth moves*! However, Galileo went further than Copernicus, who theorized that the sun is at the absolute center of the universe, and that the earth orbits about it, along with the other planets of the solar system. Galileo said there is no absolute center of the universe and that 'motion' *per se* is not an objective quality of matter.[3] Rather, it is a subjective element in its description. Thus, in Galileo's view, it is just as true to say, *from the earth's perspective*, (i.e. its frame of reference) that the sun rotates about the earth, as it is to say that, *from the sun's perspective*, the earth rotates about the sun. Indeed, it was Galileo's view that the laws of nature underlying the binding of the earth to the sun (and vice versa) are independent of the perspectives taken from which the observations of their binding ensues. This is called *Galileo's principle of relativity*.[4] It is an important precursor for *'Einstein's principle of relativity'* that underlies his theory of relativity.

In the generation that followed Galileo, Newton (who was born the same year as Galileo died, 1642) discovered the laws of motion of things (discrete masses) and the law of universal gravitation. He also perfected the means of viewing the night sky with a new, higher-resolution type of telescope — a 'reflecting telescope' — that was a collection of mirrors rather than lenses, much smaller in dimension than Galileo's lens-type telescope. Newton also did research in his attempt to understand optical phenomena. His view was a mechanistic one, where light is assumed to be a collection of particles. (*Newton's contemporaries, Hooke and Huygens, believed in a continuous wave theory of light. It was later verified that Newton's corpuscular model of light was wrong, that light is really a continuous wave phenomenon, whose underlying basis is electromagnetic radiation.*)[5]

In the 18th century, William Herschel discovered that the 'Milky Way' is not the entire universe, as Galileo believed. Herschel found that our galaxy, 'Milky Way', has a neighboring galaxy of stars, called 'Andromeda'.[6] This was discovered later on to be a member of a binary system with 'Milky Way'. Later in the history of astronomy, it was found that the universe consists of an indefinitely large number of galaxies, each containing an indefinitely large number of stars. The galaxies are not distributed in the universe homogeneously and isotropically. Rather, they cluster in certain regions and are absent in others.

With the new high-resolution instrumentation, it was found in the 20th century that most of the galaxies are pancake-shaped, with spiral arms, where the highest density of stars is toward their centers. Our sun is an average-sized star residing in one of the spiral arms of the 'Milky Way'. We also know that the galaxies rotate about an axis that is perpendicular to their planes. Some of the galaxies that do not have spiral arms have the (egg-like) shapes of ellipsoids. It is possible these are in an evolutionary stage, later to develop spiral arms. It was found that the rotations of the galaxies cannot be attributed to the Newtonian gravitational pull of the neighboring galaxies. They would not have the sufficient magnitude. It has been proposed that there is unseen *dark matter* in the regions of the galaxies (and throughout the universe) that is responsible for their rotations. The details of dark matter will be discussed in Chapter 5 (pp. 51–59).

Newton's equations of motion and his theory of universal gravitation imply all of the heavenly bodies are in stationary orbits about other bodies, just as the planets of our solar system, *from the perspective of the sun*, are in stationary orbits about the sun. This is because of the separation of the (relative) functional dependence on spatial coordinates from the functional dependence on the (absolute) time coordinates in Newton's equations of motion. The equation of motion in the spatial coordinates, in Newton's theory, predicts the elliptical orbits and constant angular momenta of a rotating body, such as the earth, relative to its center of rotation, the sun, whose center of mass resides at one of the elliptical foci. This confirms Kepler's discovery, a generation before Newton, that the orbit of Mars is elliptical about the sun, which is at one of the elliptical foci, and his generalization of this observation to a law of orbital motion of all the planets.

Is Newton's Theory an Explanation of Gravity?

One feature of Newton's theory that puzzled him was 'action at a distance'. That is, a body spontaneously acts on another body, irrespective of the magnitude of their spatial separation. He said the following about this concept: *'Action at a distance through a vacuum, without mediation of anything else by and through which their action and force may be conveyed from one to another, is to me so great an absurdity, that I believe no man who has in philosophical matters competent faculty of thinking, can ever fall into it'*. He said this in a letter to R. Bentley III, in 1693. [See A. Janiak (ed.); *Newton: Philosophical Writings* (Cambridge, 2004), p. 102. Also see M.B. Hesse, *Forces and Fields* (Nelson, 1961), p. 152]. But he excused his use of the concept by saying: 'because it works! I do not form hypotheses'. Yet, from a study of his writings, it is clear that Newton did indeed form hypotheses in physics! Because of his unhappiness with 'action at a distance', I believe Newton did not yet think that he had *explained* gravitation, though he *described* it adequately for his time.

It was not until the fruition of Einstein's theory of general relativity, three centuries later, that gravitation was explained satisfactorily. Two features that differed in Newton's versus Einstein's theories of gravitation were: (1) atomism versus the continuous field concept and (2) action at a distance versus the propagation of forces at a finite speed. A further difference was that in Newton's theory, the space parameters and the time parameters are separate from each other in the sense that the time measure is absolute — it does not change in the expression of a law of nature from one reference frame to another, but a spatial measure in one reference frame is transformed to different spatial measures in other reference frames where the law of nature is expressed. But in the theory of relativity, a spatial (or a time) measure is transformed to a mixture of a spatial and a time measure in other reference frames where the laws of nature are expressed. The time measure then becomes relative to the reference frame in which the law of nature is expressed. This is in contrast to Newton's classical theory, where the time measure is absolute, i.e. frame-independent. Thus, in relativity theory, space and time become *spacetime*. Such a language for the laws of nature then does not predict stationary orbits for the heavenly bodies.

The Expanding Universe

The lack of stationary orbits of the heavenly bodies of the universe was verified by Hubble in his discovery of the expanding universe.[7] This observation was in agreement with the prediction of no stationary orbits of matter of the universe, according to Einstein's theory. In our local domain of the solar system, at first glance there seems to be the stationary orbits of the planets relative to the sun, but Einstein's theory showed that these are really not stationary either, as exemplified in the observation of the perihelion precession of Mercury's orbit (i.e. the planet Mercury does not return to the same place relative to the sun, after each cycle of its orbital motion). Hubble's data revealed the fact that the galaxies of the universe are moving away from each other at an accelerating rate, in accordance with the Hubble law: $v = Hr$, where v is the speed of one galaxy relative to another and r is their mutual separation. This cosmological dynamics is referred to as 'the expanding universe'.

This dynamics does not mean to imply that the matter of the universe, as a whole, is expanding into empty space! For there is no empty space as a 'thing in itself'. The universe is, by definition, all there is. What is meant, physically, by the 'expansion' of the universe is that from any observer's view, the density of matter, anywhere in the universe, is decreasing with respect to his or her time measure.

The way that Hubble detected this expansion was to measure the Doppler Effect of the radiation emitted by the galaxies. If a source of radiation emission is moving away from an absorber of this radiation, its absorbed frequency components will shift toward the longer wavelength end of the spectrum — this is a 'red shift' for the visible spectrum. Thus, Hubble saw the spectrum from a more distant galaxy red shifted from a similar spectrum of a closer galaxy. This would mean that the density of matter in the observed stellar domain of the universe is decreasing in time, according to any particular observer's measurements.

If the universe is indeed monotonically expanding in this way, then if one should extrapolate to the past, the universe was increasingly dense at the earlier times. The extrapolation would then reach a limit, where the matter of the universe was maximally dense and unstable. At that point in time, according to any observer's view, the universe would have exploded, starting off the presently observed expansion. This event of the universe is commonly referred to as the 'big bang'.[8]

The question then arises: how did the matter of the universe get into this state of maximum density and instability in the first place? This is a scientific question and requires a scientific answer. To answer this question theologically by saying that this point in time was when God created the universe is a nonsequitur. It is based on religious truth, founded on irrefutable faith. That is not to say religious truth is false. Rather, it is in a different context to scientific truth. Scientific truth is in principle refutable and based on scientifically testable empirical facts and logical consistency of the basis of this theory. This difference will be discussed further in Chapter 7 (pp. 64–73).

The Oscillating Universe Cosmology

The only *scientific* answer to this question that I see is this: before the big bang event and the subsequent expansion phase of the universe, where the dominant gravitational force between galaxies was repulsive, the universe was imploding with a dominant attractive gravitational force between the matter components, contracting as a whole toward ever-increasing density. An inflection point was then reached, where the dominant attractive forces changed to dominant repulsive forces, and the contraction changed to an expansion. Eventually, when the density of the matter of the presently expanding universe will become sufficiently rarefied, another inflection point will be reached, where the dominant repulsive forces between the matter components of the expanding universe will change to a dominant attractive force. Once again, the expansion will then change to a contraction and the matter of the universe will implode until the next 'big bang', initiating the next expansion phase. The dynamics of the universe as a whole continue in cyclic fashion between expansion and contraction. This is compatible with the meaning of time in relativity theory as a non-absolute measure. That is to say, this model in cosmology rules out the concept of an absolute temporal *beginning* of the universe, from a scientific stand.

The oscillating universe cosmology follows in physics from the theory of general relativity. The terms that play the role of force in this theory depend on the geometrical functions of the curved spacetime, called 'affine connection'.[9] They are not positive-definite. Thus, under one set of physical conditions, when matter is sufficiently dense and the relative speeds of matter are close to the speed of light, the gravitational forces are repulsive, predicting that matter moves away from other matter. But when the matter becomes sufficiently rarefied in the expansion phase, the dominant repulsive forces between matter become dominantly attractive, leading to the contraction of the matter of the universe. (In the latter phase of the oscillating universe, the *redshift* of radiation observed by Hubble, implying an expansion would change to a *blueshift* — a shift of the visible spectrum toward the shorter wavelengths, as observed when the source of the radiation is approaching the absorber of this radiation, rather than receding from the absorber. This would imply a contracting universe.)

Thus, with this scenario, the universe is continually expanding and contracting — there were continual 'big bangs' in the indefinite past and will continue in the indefinite future. The time of the last 'big bang', estimated (from the Hubble law) to be about 15 billion years ago, was only the beginning of this particular cycle of the oscillating universe.

The Theory of General Relativity

The underlying physical dynamics of the universe as a whole is given by Einstein's theory of general relativity. This theory is based on a single axiom: 'The principle of relativity' (also known as the 'principle of covariance'). It is the assertion that any law of nature for any particular phenomenon, as expressed by any observer, with respect to a comparison in all possible reference frames relative to his or her own reference frame, must be in *one-to-one correspondence*. This is equivalent to saying that the laws of nature must be totally objective — independent of reference frames.

It might be argued that the theory of relativity, according to this definition, is a tautology rather than a science. For how could a law be a law, by definition, if it were not fully objective? It would be like saying: a human female is a woman. Of course this is a true statement, but it is only the definition of a word. (*It is called a 'necessary truth'*.) Nevertheless, the principle of relativity is not a tautology because it depends on two implicit assumptions that are not tautological. One is the idea that there are laws of nature in the first place. The second implicit assumption is that we can comprehend and express the laws of nature. These are not *necessary truths*. They are *contingent*, and thus qualify as scientific statements. (*This view supports Karl Popper's distinction between a necessary truth and a scientific truth.*[10])

The assertion of the existence of laws of nature corresponds to saying that for every physical effect in the world, there is an underlying physical cause. Here, there is an implication that the universe is totally ordered. It is the scientist's obligation, then, to pursue this order in terms of the cause-effect relations — the laws of nature. If, as Galileo believed, it is impossible to achieve a *complete knowledge* of the order of the universe,[11] regarding the laws for any of its phenomena, it is still the obligation of *scientists* to pursue an *increase* in their knowledge and understanding, though never expecting to reach complete comprehension.

The Role of Space and Time

The second implicit assumption is that we can comprehend and express the laws of nature, in our own language. It is the latter assumption where the space and time measures come in. They are the 'words' of a language that we invent (not the only possible language!) for facilitating an expression of the laws of nature.

It is important to recognize that the laws of nature, *per se*, are not identical with the language that we invent to express them. Nature is there, it is *of* the universe, whether or not we are there to express its characteristics. Our language, on the other hand, is invented by us, to help us express the laws of nature. Space and time are not physical things that can contract, distort, and so on. Thus, it is not correct to say that the existence of matter distorts spacetime, as a lead sphere may distort a rubber sheet that it falls into! Space and time are not more than the elements of a language we invent for the purpose of facilitating an expression of the laws of nature — such as the physical laws of the universe. This is a mathematical language that is continually perfected, perhaps for want of a better mathematical language to express the laws of nature.[12]

As we have discussed earlier, three space parameters and the time parameter in the laws of nature, according to relativity theory, do not have any objective (i.e. frame-independent) significance in themselves. What is significant as an objective language is the unification of space and time into spacetime. This means that purely space measures in the expression of a law of nature in one reference frame is a combination of space and time parameters in the expressions of the same law of nature in different reference frames. It then follows that the time measure must be expressed in the same units as the space measure. That is, in the different reference frames where the laws of nature are compared, instead of calling the time measure t seconds, t' seconds, t'' seconds. etc., they must be called ct centimeter, ct' centimeter, ct'' centimeter, etc. where c, as a conversion factor, must be frame-independent, with the dimension of centimeter/second — the dimension of a speed. Thus, the principle of relativity predicts there must be an invariant speed associated with the time measure. It turns out, in looking at one particular law of nature — the Maxwell field equations that underlie electromagnetism — that c is the invariant speed of light in a vacuum. (*In the initial stages of relativity theory, Einstein said that there were two independent axioms that underlie this theory: (1) the principle of relativity and (2) the invariance of the speed of*

*light. We see here that the latter is not an independent axiom —
it follows logically from the principle of relativity.)*

The spacetime language forms a continuous set. The principle of
relativity then requires that the transformations of the expressions
of the laws of nature, such as the laws of cosmology, in one
reference frame, to a *continuously connected* reference frame, must
leave the form of the law unchanged. (It is called 'covariance'.) This
requirement implies that the laws themselves must be *continuous
field equations* (as is the Maxwell formulation for electromagnetism)
and that the solutions of these laws are *continuous fields* (everywhere).
One further requirement of the transformations of relativity theory, according to the principle of relativity, is that discovered by
Emma Noether, called *'Noether's theorem'*.[13] It is that a necessary
and sufficient condition for the laws of conservation (of energy,
momentum and angular momentum) to be included with the other
laws of nature, in the special relativity limit of the theory, is that
the transformations must not only be continuous, but also analytic.
That is, their derivatives must exist to all orders. Singularities are
then automatically excluded. The set of continuous, analytic transformations, everywhere, forms a *Lie group*. In the case of the theory of
general relativity, this is the 'Einstein group'. In the special relativity
limit of this theory, this is the 'Poincaré group'. These Lie groups
underlie the algebraic logic of the theory of relativity. The functions
that transform in this way, to maintain the principle of relativity,
are called 'regular', indicating the inadmissibility of singularities
anywhere in the universe! This would include a rejection of the
'black hole' (as it is commonly understood in physics today), and
the singularity of the 'big bang' in cosmology. It is a condition on the
solutions of general relativity, emphasized by Einstein throughout
his lifelong pursuit of this theory. That is, Einstein would not have
accepted either of these singularities, commonly discussed today,
as realities![14] (*In personal discussions with Prof. Nathan Rosen, one of
Einstein's close collaborators in the 1930s, he acknowledged the validity of
this statement.*)

The Lie group of the theory of general relativity, the Einstein
group, has 16 essential parameters; these are the derivatives of

the four spacetime coordinates of one (primed) reference frame where the laws are expressed, with respect to those of a different (unprimed) spacetime reference frame, $\partial x^{\mu'}/\partial x^{\nu}$, where μ and ν refer to the four space and time coordinates in each reference frame. The significance of the number 16 of essential parameters of the Lie group of general relativity theory is that there must be 16 independent field equations to prescribe the spacetime.[15] What is interesting here is that Einstein already showed that there are (at least) 10 independent equations (the 'Einstein field equations'), with 10 solutions, that underlie the gravitational phenomenon, and there are 6 independent solutions of the laws of electromagnetism — the three components of the electric field and the three components of the magnetic field. It has been shown that the Lie group of general relativity then implies a truly unified field theory, with a 16-component metrical field, where gravity and electromagnetism are unified in terms of a single field of force. This new 16-component metrical field of general relativity theory is a four-vector field $q^{\mu}(x)$, in which each of the four components is a quaternion rather than a real number field.[16] *The new quaternion formulation of general relativity theory implies a cosmology in which the expansions and contractions of the oscillating model of the universe, as a whole, are spiral rather than isotropic, and where the Hubble law is predicted as a first approximation. This is discussed in Chapter 4 and in Ref. 17.*

Geometry and Matter

Einstein started his theory of general relativity with the idea that the variability of the matter content of the universe implies a variability of the coefficients of the metric tensor that underlies the geometry in the language of spacetime. This is tied to the idea that the spacetime is not more than a language that *reflects* the physical properties of the matter content of the universe. Analogous to the logic of ordinary verbal language, this language, in turn, has a geometrical logic and an algebraic logic. The algebraic logic is in terms of the underlying symmetry group of the theory, as discussed above. The geometrical logic is expressed in terms of the invariant differential metric of the spacetime:

$$ds^2 = g^{\mu\nu}(x)dx_\mu dx_\nu = ds'^2,$$

where summation is implied over the subscripts and superscripts, and $\mu, \nu = 0, 1, 2, 3$ are the temporal (0) and spatial (1, 2, 3) coordinates. $g^{\mu\nu}(x)$ is the 'metric tensor'. It is, in Einstein's nonsingular field theory, a 'regular function', *everywhere*, and a second-rank, symmetric, 10-component tensor. Its continuous variability in the spacetime x is a *reflection* of the continuous variability of the matter fields of the spacetime. The idea is then that the spacetime transformations that keep the invariance of $ds^2 = ds'^2$ are those that transform *any* law of nature covariantly — i.e. preserving its form in all (continuously connected) reference frames. This is a *Riemannian geometrical system*.

The geodesic of the spacetime — the path of minimum separation between any two of its points — is determined by minimizing the invariant integral that is the path length $\int ds$. The physical significance of the geodesic is that it is the natural path of an unobstructed body. In Galileo's classical view, the geodesic is a straight line; the family of such straight lines is a 'flat spacetime'. *Galileo's principle of inertia* then asserts the natural path of an unobstructed body is a straight line. In contrast, the geodesic of the Riemannian spacetime is a curve. Thus, the natural path of an unobstructed body in such a spacetime is a curve. The family of such curves is a 'curved spacetime'. The variables of this curve, that gives it its structure, are determined by the variability of the matter of the spacetime. As the matter of the system is then continuously depleted, the curved spacetime approaches a flat spacetime. In this limit of a perfect vacuum, *everywhere*, the values of the metric tensor become:

$$g^{00} \to 1, \quad g^{kk} \to -1, \quad g^{\mu\neq\nu} \to 0, \quad (k = 1, 2, 3)$$

so that, for a perfect vacuum, *everywhere*, the invariant metric becomes:

$$ds^2 = (dx_0)^2 - (dx_1)^2 - (dx_2)^2 - (dx_3)^2 = ds'^2$$

This is indeed the invariant metric of special relativity theory — the 'Lorentz metric'. Thus, the theory of special relativity, discovered originally to preserve the forms of the laws of nature in inertial fames of reference (i.e. frames in constant, rectilinear speed relative to each other) really only applies to the ideal case of the perfect vacuum. Where its formalism 'works' in physics, for a material medium (e.g. in its prediction of the energy- mass relation, $E = mc^2$, the dynamics of elementary particles, etc.) must then be a mathematical approximation for the metric of general relativity. This assumes the curvature of spacetime, in a small region, may be *approximated* by a flat spacetime, tangent to the curved spacetime where it is applied.

The idea of the theory of general relativity, as discussed earlier, is that the geometry of spacetime is to *reflect* the matter content of the closed system, in principle the universe. For example, the existence of the sun is *reflected* in the curved spacetime geometry in the vicinity of the sun, causing the trajectory of a beam of starlight to bend as it passes this region of space. This bending was predicted qualitatively and quantitatively by Einstein and it was then observed in agreement with his theory of general relativity. It is a gravitational effect that was not predicted by the earlier Newtonian theory of universal gravitation. Along with other gravitational effects not predicted by the classical theory, as well as giving back the equations of the classical theory *as an approximation*, Einstein's theory of general relativity superseded Newton's theory of universal gravitation, as a true *explanation* of gravity. Thus, Einstein discovered the field equations that predict the phenomenon of gravity. It relates the geometry of spacetime, in terms of the Riemannian metric tensor, $g^{\mu\nu}(x)$ and its changes in spacetime to the matter field variables of the closed system, chosen to be the energy-momentum tensor of its material content.

Generalization of Einstein's Field Equations

Einstein's field equations, in $g^{\mu\nu}(x)$, are 10 independent, nonlinear differential equations. But they are too few in number, for this reason: the symmetry group of general relativity theory — the 'Einstein group' — is the group of transformations that defines the invariance of $ds^2 = g^{\mu\nu}dx_\mu dx_\nu = ds'^2$. This is a *Lie group*, a group of continuous, analytic transformations that preserves the forms of the laws of nature under the changes from one frame of reference to any continuously connected frame of reference. This preservation is the requirement of the underlying principle of relativity. The number of *essential parameters* of the Lie group, $\partial x^{\mu'}/\partial x^\nu$, is 16 in number. It implies that the most general form of the field equations, subject to the principle of relativity, must be 16 in number rather than 10.

Why are Einstein's equations 10 in number rather than 16? It is because the form of these equations is more symmetric than they need be, in accordance with the (16-parameter) Einstein group. They are covariant (form-invariant) with respect to the continuous transformations, as they must be. But they are also covariant with respect to the discrete reflections in space and time, which is not required. By lifting the space and time reflection transformations, Einstein's equations thereby *factorize* to 16 independent equations, as it is shown in Chapter 3 (pp. 27–36) and in Ref. 18.

It is important to note that the invariant differential metric of the spacetime is not ds^2, it is ds. How does one take the square root of ds^2? It is usually stated that its square root is $\pm ds$, and the minus sign is simply thrown away! But this is not valid, as the square root is double-valued at all points of the spacetime, i.e. there is an ambiguity in the sign of this term, *everywhere*.

The answer to this question comes from the fact that the irreducible representations of the Einstein group of general relativity obey the algebra of quaternions (as well as the irreducible representations of the Poincaré group, of special relativity). The quaternion is a generalization of the complex number in 2-dimensional space, whose basis elements are 1 and i ($= \sqrt{-1}$), forming the complex function, $f(z = x + iy) = u(x,y) + iv(x,y)$.[19] The basis elements $(1, i)$ of the complex number in two dimensions generalize to the four basis elements, σ^0, σ^k, (where $k = 1, 2, 3$); these are the unit two dimensional matrix and the three Pauli matrices. Thus, the quaternion has the *form* in the 4-dimensional space

$$q(x) = \sigma^0 x_0 + \sigma^1 x_1 + \sigma^2 x_2 + \sigma^3 x_3$$

We then define the quaternion four-vector $q^\mu(x)$, with the differential invariant quaternion of general relativity,

$$ds = q^\mu dx_\mu$$

Thus, there is no ambiguity here as to the sign of the invariant ds. It is important, and clear from this form of the quaternion, in terms of two-dimensional matrices that the product of two quaternions is not commutative under multiplication, i.e. $q_1 q_2 \neq q_2 q_1$.

Each of the four components of the four-vector q^μ is quaternion-valued and thus has four components. Thus, this quaternion metrical field has $4 \times 4 = 16$ components. This is a unique expression for the invariant $ds = q^\mu dx_\mu$. Further, the quaternion metric field transforms as a second rank spinor, that is, q^μ transforms as $(\psi \times \psi^*)^\mu$. Thus, the basis elements of the quaternion are two-component spinor variables ψ.

We see here that any law of nature, *whether in particle physics or in cosmology — the physics of the universe —* that is compatible with the symmetry required by relativity theory, in special or general relativity, must be in terms of spinor and quaternion variables. This is a requirement of the algebraic logic — the group structure — of the theory of relativity. It is the reason why Dirac's special relativistic theory of wave mechanics led to spinor degrees of freedom in the description of the electron (and a quaternion operator to determine these solutions). That is, the spin degrees of freedom in Dirac's electron equation are not a consequence of quantum mechanics, *per se*, as many have claimed! It is a consequence of the symmetry imposed by the theory of relativity.

The correspondence of the quaternion metric field and the metric tensor of Einstein's formulation is then in terms of the product of ds and its quaternion conjugate ds^* (corresponding to its time (or space) reflection):

$$ds\, ds^* \approx -(1/2)(q^\mu q^{\nu*} + q^\nu q^{\mu*})dx_\mu dx_\nu = g^{\mu\nu}dx_\mu dx_\nu$$

Thus, the quaternion formulation in general relativity, $ds = q^\mu dx_\mu$, is a factorization of the metric tensor formulation of the standard Einstein theory.

Similar to Einstein's derivation of the field equations in $g^{\mu\nu}$ from a variational principle, the factorized field equations in q^{μ} may be derived from a variational principle. These derivations are shown in Ref. 16. The quaternion form yields 16 field relations that replace the 10 relations of Einstein's symmetric tensor form of his field equations in general relativity. In addition to explaining gravity, as will be described in Chapter 3 (pp. 27–36), the quaternion form predicts new physical effects in the cosmological problem of the universe as a whole. One important new feature is the torsion of spacetime. It predicts, as examples, the rotation of the galaxies and the Faraday Effect regarding the propagation of cosmic electromagnetic radiation, i.e. the rotation of the plane of polarization of this radiation as it propagates throughout the universe. Both are observed astrophysical effects, not predicted by the standard Einstein tensor formulation. A further prediction is an anisotropic expansion and contraction of the universe, in a spiral fashion. Another important difference is that the geodesic equation, that prescribes a natural motion along a curve of an unobstructed body, has a quaternion form. That is, to prescribe the motion of a body along a trajectory, parameterized by the time measure, one must have four parameters, rather than one, to prescribe the time change, as a body moves from one spatial location along its trajectory to another. This is the generalization of the time parameter in physics theorized by William Hamilton, from his discovery of the quaternion algebra in the 19th century.

A Unified Field Theory

By iterating the 16 field equations in q^{μ} with the conjugated solution $q^{\nu*}$ on the left, and iterating the conjugated (i.e. reflected) equation in $q^{\nu*}$ on the right with q^{μ}, second-rank tensor equations are generated. Adding these two iterated equations generates a symmetric second-rank tensor equation (10 components) that we will see is in one-to-one correspondence with Einstein's original tensor equations (Chapter 3), thus explaining gravity. Subtracting these equations generates an antisymmetric second-rank tensor equation that can then be put into one-to-one correspondence with Maxwell's equations, thus explaining electromagnetism. Thus, the original 16 quaternion metrical field equations break up into 10 equations that explain gravity and 6 equations that explain electromagnetism — *this is the unified field theory that was sought by Einstein.* In this field theory, in general relativity both physical phenomena are incorporated in the single, 16-component quaternion field q^μ. It is this formal, generalized expression of general relativity that generates a new cosmology,[17] relating to the physics of the universe. The dynamics is an oscillating universe, between expansions and contractions, in spiral configuration. As will be discussed in Chapter 4 (pp. 37–50), the Hubble law is an approximation for this dynamics, over sufficiently short times in the expansion phases of the universe.

In the next chapter, we will discuss the physics and outline the formal development of Einstein's tensor form of general relativity theory as a language of cosmology.

A Language of Cosmology: The Mathematical Basis of General Relativity

Introduction

The basis of the theory of general relativity, as underlying the physics of the cosmos, was discussed in Chapter 1. In this chapter we will outline the mathematical derivation of this formalism, as it was shown in Einstein's study. A generalization will then be shown in the next chapter from Einstein's tensor formulation to the quaternion formulation. This leads from the 10 components of the symmetric tensor solutions to 16 independent components of the quaternion solution for the metric of spacetime.[16] In Chapter 3, we will utilize the quaternion formulation to demonstrate (1) a unification of electromagnetism with gravity, (2) an oscillating universe cosmology, and (3) as a first approximation, the Hubble law, indicating the expansion of the universe. In Chapter 4, the spiral configuration of the expanding and contracting universe will be demonstrated explicitly. As a bonus, it will be shown in Chapter 5 that the spiraling matter of the oscillating universe provides a mechanism for the creation of magnetic fields of opposite polarity that separate out matter from antimatter of the universe at the beginning of each expansion phase of the universe.

Einstein's Tensor Formulation[20]

The invariant differential metric with the Riemannian geometry of spacetime is:

$$ds^2 = g^{\mu\nu}(x)dx_\mu dx_\nu = ds'^2 = g^{\mu\nu'}(x')dx'_\mu dx'_\nu \qquad (2.1)$$

where the summation convention is assumed over the space and time coordinates, $\mu, \nu = 1, 2, 3, 0$.

The metric tensor $g^{\mu\nu}(x)$ is a regular (continuous and analytic) field, *everywhere*, as has been argued in the preceding chapter. The geodesics of this spacetime, i.e. the paths wherein the distance between their points is a minimum, are *variable curves*, rather than the straight lines of a Euclidean spacetime.

The physical significance of the geodesic is that it is the natural path of an unobstructed body. To remove the body from this path would require the input of external energy. The family of such curved geodesics is called a 'curved spacetime'. In the limit, as the matter of a closed system (in principle, the universe) is depleted toward a vacuum, *everywhere*, the curved spacetime approaches a 'flat spacetime'. In this limit, the invariant Riemannian metric (2.1) would become the invariant Euclidean (Lorentz) metric of special relativity,

$$ds^2 = (dx^0)^2 - dr^2 = ds'^2 = (dx^0)'^2 - dr'^2 \qquad (2.1')$$

The principle of relativity then requires that the same spacetime transformations that leave ds^2 invariant (Eq. (2.1) or (2.1')) must leave all of the laws of nature *covariant* (i.e. frame- independent).

The Riemann Curvature Tensor

As we have discussed earlier, it is the thesis of the theory of general relativity that the geometry of spacetime is to *reflect* the matter content of the universe. Einstein discovered the field equations that express this reflection of matter in geometry. The left-hand side of his equations is: $G_{\mu\nu} = R_{\mu\nu} - (1/2)g_{\mu\nu}R$. This is called 'the Einstein Tensor'. As it will be shown below, the Ricci tensor, $R_{\mu\nu}$, also entails the geometrical field, the metric tensor $g_{\mu\nu}$. The right-hand side of Einstein's equations is chosen to be the energy-momentum tensor $T_{\mu\nu}$, representing the material content of the universe. The tensor $R_{\mu\nu}$, and the 'scalar curvature', R, will be defined explicitly below. The origin of the left-hand side of Einstein's equations, the Einstein tensor, $G_{\mu\nu}$, is in the Riemannian curvature tensor $R^{\lambda}_{\mu\nu\varrho}$, that is defined in terms of a difference of second covariant derivatives of a vector field, V_{μ}, as follows:

$$V_{\mu;\nu;\varrho} - V_{\mu;\varrho;\nu} = R^{\lambda}_{\mu\nu\varrho} V_{\lambda} \qquad (2.2)$$

As the flat spacetime is asymptotically approached (i.e. toward a perfect vacuum *everywhere*), $R^{\lambda}_{\mu\nu\varrho} \to 0$ and the covariant, second derivatives of the vector field V_{μ} become the ordinary second derivatives of this field, which do not depend on the order of differentiation.

The Riemann curvature tensor $R^{\lambda}_{\mu\nu\varrho}$ is derived from the covariant derivatives of the four-vector field V_{ν} as follows:

$$V_{\nu;\lambda} = \partial_{\lambda} V_{\nu} - \Gamma^{\varrho}_{\lambda\nu} V_{\varrho} \qquad (2.3)$$

where the coefficients of the 'affine connection' are[16]:

$$\Gamma^{\varrho}_{\mu\alpha} = (1/2)g^{\varrho\lambda}(\partial_{\mu}g_{\lambda\alpha} + \partial_{\alpha}g_{\mu\lambda} - \partial_{\lambda}g_{\alpha\mu}) \qquad (2.4)$$

The combination of eqs. (2.2) and (2.3) then gives the Riemannian curvature tensor in terms of the affine connection as follows:

$$R^{\lambda}_{\mu\nu\varrho} = \partial_{\nu}\Gamma^{\lambda}_{\mu\varrho} - \partial_{\varrho}\Gamma^{\lambda}_{\mu\nu} + \Gamma^{\lambda}_{\alpha\nu}\Gamma^{\alpha}_{\mu\varrho} - \Gamma^{\lambda}_{\alpha\varrho}\Gamma^{\alpha}_{\mu\nu} \qquad (2.5)$$

The Ricci tensor, in turn, is a contraction of the Riemann curvature tensor, as follows:

$$R_{\mu\nu} = R^{\lambda}_{\mu\nu\lambda} = \partial_{\nu}\Gamma^{\lambda}_{\mu\lambda} - \partial_{\lambda}\Gamma^{\lambda}_{\mu\nu} + \Gamma^{\lambda}_{\alpha\nu}\Gamma^{\alpha}_{\mu\lambda} - \Gamma^{\lambda}_{\alpha\lambda}\Gamma^{\alpha}_{\mu\nu} \qquad (2.6)$$

It is readily shown[16] that the Ricci tensor is symmetric, $R_{\mu\nu} = R_{\nu\mu}$. The scalar curvature is, by definition, the contraction

$$R = g^{\mu\nu}R_{\mu\nu} \qquad (2.7)$$

Because the covariant divergence of the energy momentum tensor vanishes, $T^{;\nu}_{\mu\nu} = 0$ (the flat space limit of this law is $\partial^{\nu}T_{\mu\nu} = 0$ — the conservation law of energy and momentum), the covariant divergence of the left-hand side of Einstein's equations, that is, of the Einstein tensor $G_{\mu\nu}$, must vanish. Analysis indicates[16] that $R^{;\nu}_{\mu\nu} = (1/2)(g_{\mu\nu}R)^{;\nu}$. Thus, $G^{;\nu}_{\mu\nu} = (R_{\mu\nu} - (1/2)g_{\mu\nu}R)^{;\nu} = 0$, so that we have the Einstein field equation:

$$G_{\mu\nu} \equiv R_{\mu\nu} - (1/2)g_{\mu\nu}R = kT_{\mu\nu} \tag{2.8}$$

wherein the covariant divergence of both sides of this equation vanish.

The derivation of Einstein's field equations (2.8) from a variational principle is demonstrated in Reference 16.

Summing up, the field equations (2.8) with (2.4), (2.5) and (2.6) are 10 nonlinear differential equations, whose solutions are the components of the metric tensor $g_{\mu\nu}$. Once these solutions are determined, they may be inserted into the geodesic equation, shown below, to determine the path of an unobstructed body in the curved spacetime.

The Geodesic Equation

The geodesic equation follows from a minimization of the indefinite line integral $\delta \int ds = 0$. This yields the geodesic 'equation of motion'[21]:

$$d^2x^{\varrho}/ds^2 + \Gamma^{\varrho}_{\mu\nu}(dx^{\mu}/ds)(dx^{\nu}/ds) = 0 \tag{2.9}$$

Thus, with the affine connection coefficients in terms of the metric tensor (2.4), the Ricci tensor (2.6) and the scalar curvature R (2.7), the solutions $g_{\mu\nu}$ of Eq. (2.8) in Eq. (2.9) yield the geodesic path of a body in the curved spacetime. With the foregoing, we then have the apparatus to determine the explicit effects of gravitational phenomena.

The Vacuum Equation

The actual numerical successes of the formalism of general relativity, thus far, come from the assumption of a vacuum, $T_{\mu\nu} = 0$, *everywhere*, yielding the nonlinear differential equation that is the 'vacuum equation':

$$R_{\mu\nu} = 0 \tag{2.10}$$

The assumption made here is that $T_{\mu\nu} \neq 0$ inside of the surface of a star, yielding the non-homogeneous equation (2.8) there, but that it is zero outside of the star, yielding the 'outside' homogeneous equation (2.10). One then *adds* the two solutions to yield the general solution, and match the solution of the homogeneous equation to the solution of the non-homogeneous equation at the surface of the star to determine the integration constants, as it is done in classical (linear) physics. However, this is not valid here, in principle, since all fields in general relativity solve nonlinear equations, thus they are not additive *anywhere*, resulting in a different solution. There cannot be a sharp cut-off, where the *continuous field* $T_{\mu\nu}$ changes from a nonzero value, as inside of a material medium (a star), to a zero value, outside of the star. This is because of the nonlinear mathematical structure of the field theory. There is, in fact, no objective 'inside' and 'outside' of a body, such as a star. That is, the general solution of a nonlinear differential equation is *not* the sum of a homogeneous vacuum solution, where there is zero on the right (the outside of the star) and a solution for non-zero on the right (the inside of the star). It then follows that in general relativity we must interpret the vacuum equation (2.10) as an *approximation* for the actual field equation of general relativity in Einstein's tensor formulation (2.8).

The Schwarzschild Solution[22]

With the definition of the Ricci tensor in (2.6) and the affine connection in (2.4), the nonlinear differential equation for the vacuum (2.10) in $g_{\mu\nu}$ was shown by Schwarzschild to yield the solution:

$$ds^2 = (1 - 2\alpha/r)(dx^0)^2 - dr^2/(1 - 2\alpha/r) - r^2 d\Omega^2 \tag{2.11}$$

where α is a constant of the integration (with the dimension of length) and $d\Omega$ is the differential solid angle. Comparison with the limit, in which the Newtonian theory of gravity appears as an approximation, yields: $\alpha = GM/c^2$, where G is the gravitational constant in Newton's force law: $F = Gm_1m_2/r^2$, (m_1 and m_2 are the gravitationally interacting masses and r is their mutual separation), and M is the mass of the body that gives rise to the metric (2.11). We see there is a singularity in this solution at the radial distance $r = 2\alpha$. (For the sun, this is the order of a few kilometers from its center — well inside the body of the sun).

In view of Einstein's admission of only nonsingular (regular) solutions of general relativity, this singularity is not real. When the vacuum equation (2.10) is extended to the full form of Einstein's field equations (2.8), according to their meaning in Einstein's view, the apparent singularity washes away.

The black hole

Nevertheless, for those who see (2.11) as an exact solution of general relativity, there is a peculiar type of star predicted, if the matter of the sun could be condensed to a sphere with the Schwarzschild radius, $r = 2\alpha$ (a diameter of a few kilometers for the sun) without blowing apart before reaching this dimension and enormous density! The singularity at this 'event horizon' would prevent any signal from propagating away from the star. This type of star is called a 'black hole'.[23]

As mentioned above, this singular solution is inadmissible in view of our acceptance of only regular (nonsingular) solutions of the equations of general relativity. Still, there is a possibility of a 'black hole' in a different context. If a star is sufficiently dense that the geodesics associated with it are *closed*, then all emissions from this star, including light and gravitation, that would be propagated away from the star would then return to it, along the closed geodesic, and thus reabsorbed by the same star. It would then be black to any outside observer. Such a star may then also be called a 'black hole'. *The possibility of the existence of such a star would depend on the existence of 'stable solutions' associated with the closed geodesics. This has not yet been established, theoretically, nor is there yet any conclusive experimental proof for the existence of a black hole.*

The Crucial Tests of General Relativity

The Schwarzschild solution (2.11) of the vacuum equation (2.10) leads to the three crucial tests of the theory of general relativity. The first is the perihelion precession of Mercury's orbit.

This effect was observed in the 19th century, but unexplained with classical physics. It was predicted, both qualitatively and quantitatively, correctly in the 20th century by general relativity.

The second effect was the bending of a beam of starlight as it propagates past the rim of the sun. This was observed in 1919 by a group of astronomers, headed by A. Eddington, from Cambridge University. Their observations were in agreement with the predictions of general relativity, qualitatively and quantitatively.

The third crucial test was the gravitational red shift. This is a shift of frequencies of emitted radiation in the visible spectrum toward the red end of the spectrum, as the potential energy environment of the emitter is increased. This was observed empirically in the 1950s by R.V. Pound and his co-workers at Harvard University. It was in qualitative and quantitative agreement with the prediction of general relativity.[24]

In addition, (2.10) also predicts, in a linear approximation, Newton's gravitational equation. The foregoing then constitutes the success of Einstein's theory of general relativity to supersede Newton's theory, to explain the phenomenon of gravity.[25]

It was mentioned in the preceding chapter that, in the 17th century, Newton himself was not satisfied that his theory was indeed an *explanation* of the phenomenon of gravity, though he understood it as an adequate *description*. His problem was with the role of 'action-at-a-distance' in the theory.

Still, Newton was willing to use the concept because it 'worked'. He said that he did not form hypotheses. For this reason, Newton did not believe that his theory *explained* the phenomenon of gravity, though he did believe that it correctly *described* it, in his day. It was not, indeed, *explained* until the phenomenon of gravity was correctly predicted by Einstein's theory of general relativity in the early part of the 20th century.

On the use of the vacuum equation (2.10) to accurately describe gravity, it is not to be interpreted, according to Einstein's meaning, as an exact form of the theory of general relativity. For, as we have discussed earlier, the geometry of spacetime, expressed in terms of the metric tensor, is to *reflect* the continuous matter content of the physical system; in principle, the universe. It then follows, logically, that if there would be no matter content of the universe, that is, if the universe were a perfect vacuum, *everywhere*, there would be no variable geometry to reflect this vacuum except for the flat spacetime, with the Lorentzian metric, $g^{00} = 1$, $g^{11} = g^{22} = g^{33} = -1$, and $g^{\mu \neq \nu} = 0$. In this case, $ds^2 = (dx^0)^2 - dr^2$. This is the invariant metric of the theory of special relativity theory. We see, at this juncture, that the theory of special relativity is only an exact theory for the ideal case where the entire universe would be a perfect vacuum! Yet, it is a valid *approximation* for the formalism of general relativity under special physical conditions (as in its use in particle physics), when matter is sufficiently rarefied and it does not move relative to other matter at speeds close to the speed of light. *Present-day experiments at the laboratory, CERN, in Switzerland, using the 'supercollider', are attempting to duplicate the quantities of matter density and energy transfer near the time of the (last) 'big bang'. The results of this experimentation may indeed lead to a refutation of the use of special relativity and the need for the mathematical formulation of general relativity to explain these data.*

The Logic of the Spacetime Language

The logic of the spacetime language of general relativity is in two parts: geometry and algebra. The geometrical logic relates points to points and lines in the sense of congruence, parallelism, etc. The algebraic logic relates points to points and lines in the sense of rules of combination, associativity, commutativity, etc. The algebraic logic is expressed most compactly in terms of a symmetry group. (*Modern Mathematics has shown that all of the theorems of geometry and algebra may be merged into a common set of theorems. However, for our purposes, we will discuss them separately.*)

The symmetry group of general relativity theory — the 'Einstein group' — is a set of *continuous transformations* that leave the Riemannian metric (2.1) invariant. Requiring further, regular solutions, these transformations must not only be continuous, but also analytic. The origin of this requirement is Noether's theorem.[13] It requires that a necessary and sufficient condition for the incorporation of the conservation laws — of energy, momentum and angular momentum in the special relativity limit of the field theory — is the analyticity of the field variables and their transformations. That is, the derivatives of the space and time coordinates of one reference frame with respect to the space and time coordinates of any continuously connected reference frame must exist to all orders. All such transformations are then nonsingular. Thus, the Einstein group is, by definition, a 'Lie group'. The number of *essential parameters* of this group is 16. These are the derivatives of one set of spacetime coordinates, of one reference frame, with respect to another set of spacetime coordinates, of a different (continuously connected) reference frame, $\partial x^{\mu'}/\partial x^{\nu}$.

Why, then, are Einstein's equations (2.8), 10 in number rather than 16? It is because they are more symmetric than they need be. These field equations are covariant with respect to the continuous changes in space and time, as they are required to be. But they are also symmetric with respect to the discrete reflections in space and time, which is not a requirement. Indeed, the Einstein group is a *continuous group*, without any discrete reflection transformations.

We will see in the next chapter that by dropping the reflection symmetry elements from the Einstein equations (2.8), they factorize to a new form wherein the symmetric tensor field variables $g^{\mu\nu}$ (10 components) are replaced with the quaternion field q^{μ}. This is a four-vector field, but each of the four components is quaternion-valued, with four independent components, rather than real-number-valued. Thus, this metric field has $4 \times 4 = 16$ components. This quaternion field equation for general relativity is then covariant under the continuous transformations in space and time but it is not covariant with respect to the discrete reflections in space and time.

3

A Unified Field Theory
in General Relativity:
Extension from the Tensor
to the Quaternion Language

Introduction

Einstein made the following comment about the success of his tensor formulation of general relativity[25]:

'Not for a moment, of course, did I doubt that this formulation was merely a makeshift in order to give the general principle of relativity a preliminary closed expression. For it was nothing more than a theory of the gravitational field, which was somewhat artificially isolated from a total field of as yet unknown structure'.

He then went on to say:

'To remain with the narrower group and at the same time to have the relativity theory of gravitation based upon the more complicated tensor structure implies a naïve consequence'.

If Einstein's tensor expression of general relativity (2.8) is only 'preliminary', what is its final formulation? I believe that the answer comes from Einstein's second comment, requiring a closer look at the symmetry group of the theory. As we noted in Chapter 2, the question is then: Why does Einstein's tensor formulation lead to

10 equations? The answer is that Einstein's equations (2.8) are more symmetric than they need to be.

The Einstein group is the transformation group that underlies the covariance of the laws of nature in general relativity. It is a 16-parameter Lie group. The 16 essential parameters of this group are the derivatives $\partial x^{\mu\prime}/\partial x^{\nu}$ of one set of spacetime coordinates, in one reference frame, with respect to another set of continuously connected spacetime coordinates of a different reference frame. The Einstein field equations (2.8) are covariant with respect to the continuous transformations in spacetime, as required. But they are also covariant with respect to the discrete reflections in spacetime, $x^{\mu} \rightarrow -x^{\mu}$, which is not required. Thus, the symmetry of Einstein's tensor field equations (2.8) is reducible. That is to say, Einstein's tensor field equations (2.8) are not the most general expression of his theory.

By lifting the reflection symmetry from the covariance group of Einstein's 10 tensor equations, they factorize to a set of 16 equations. These are in terms of the quaternion field that is covariant with respect to the continuous transformations, but not with respect to the discrete reflections in space and time.

Factorization of Einstein's Tensor Field Equations[26]

How does this factorization come about? It follows from the fact that the irreducible representations of the Einstein group obey the algebra of quaternions. These behave as second-rank spinors of the form $\Psi \times \Psi^{*}$, where Ψ is a two-component spinor variable.

We start with the Riemann invariant of the spacetime, the squared differential interval $ds^2 = g^{\mu\nu}dx_{\mu}dx_{\nu}$. However, what we require at the outset is the invariant interval ds, not its square ds^2. What is conventionally done is to take its square root, giving $ds = \pm\sqrt{g^{\mu\nu}dx_{\mu}dx_{\nu}}$ and then to reject the minus sign. But this is not valid since this expression of ds is double-valued, everywhere. The ambiguity of the sign remains.

What is done to factorize the squared differential ds^2 is to recognize the quaternion structure of the irreducible representations of the Einstein group and take the following form for the differential invariant metric:

$$ds = q^\mu(x)dx_\mu \tag{3.1}$$

The latter is a sum of four quaternions, therefore this form of ds is quaternion-valued, rather than real-number valued.

The quaternion metric field q^μ is a four-vector in configuration space, but each of its four components is a quaternion. Thus, q^μ has $4 \times 4 = 16$ independent components. It is the metrical field in general relativity that replaces the metric tensor $g^{\mu\nu}$. The differential metric (3.1) is then unique — there is no ambiguity in its expression. It is indeed the 'square root' of the Riemannian squared metric ds^2. The correspondence with the metric tensor is as follows:

$$ds\,ds^* = -(1/2)(q^\mu q^{\nu*} + q^\nu q^{\mu*})dx_\mu dx_\nu = g^{\mu\nu}dx_\mu dx_\nu \tag{3.2}$$

where the asterisk denotes the quaternion conjugate — its reflection in time (or, in a different convention, in space).

The $-1/2$ is chosen for normalization purposes. (It is important to note that the quaternions are not commutative under multiplication, i.e. $q_a q_b \neq q_b q_a$.)

The Riemann Curvature Tensor in Quaternion Form

The next step, then, is to express the tensor, $R^\rho_{\mu\nu\lambda}$ (the Riemannian curvature tensor), $R_{\mu\nu}$ (the Ricci tensor) and R (the scalar curvature) in quaternion form. Once this is done, we will take the Lagrangian density to be the scalar $R\sqrt{-g}$. The minimization of the action function $\delta S = \delta \int R\sqrt{-g}\,d^4x = 0$ then yields the metrical field equations in quaternion form.

The root of the language of the curved spacetime is the Riemannian curvature tensor $R^\rho_{\mu\nu\lambda}$. As we discussed in the preceding chapter, it is defined in terms of the difference of second-order derivatives of a vector field $V_\mu(x)$ in the curved spacetime, as follows:

$$V_{\mu;\nu;\lambda} - V_{\mu;\lambda;\nu} = R^{\rho}_{\mu\nu\lambda} V_{\rho}$$

where the semicolons denote the covariant derivatives. The covariant derivative is defined in (2.3) and the affine connection in (2.4). As the spacetime asymptotically becomes flat (corresponding to a depletion of all matter), the Riemann curvature tensor vanishes and the covariant derivatives become the ordinary derivatives that are independent of the order of differentiation. Equation (2.5) shows the relation of the Riemann curvature tensor to the affine connection components.

Spin-affine connection

In the quaternion calculus, it is first necessary to define the 'spin-affine connection'. Just as the covariant derivative of a vector field requires the addition of an affine connection term to be integrable in a curved spacetime, so the covariant derivative of a spinor variable requires the addition of a spin-affine connection term to be integrable in a curved spacetime. The covariant derivative of a two-component spinor is then:

$$\Psi_{;\mu} = \partial_{\mu}\Psi + \Omega_{\mu}\Psi \tag{3.3}$$

where the spin-affine connection is[27]:

$$\Omega_{\mu} = (1/4)(\partial_{\mu}q^{\rho} + \Gamma^{\rho}_{\tau\mu}q^{\tau})q^{*}_{\rho} \tag{3.4}$$

It follows from the quaternion invariance:

$$q^{\mu}q^{*}_{\mu} = \text{invariant} \tag{3.5}$$

that $q^{\mu}_{;\lambda} = 0$ and $q_{\mu;\lambda} = 0$. Thus, the second covariant derivatives of the quaternion variables must vanish. Taking into account that q_{μ} is a second-rank spinor of the type $(\Psi \times \Psi^{*})_{\mu}$ and that it is a four vector in configuration space, it follows that:

$$0 = q_{\mu;\rho;\lambda} - q_{\mu;\lambda;\rho} = [(\Psi_{\alpha;\rho;\lambda} - \Psi_{\alpha;\lambda;\rho})\Psi^{*}_{\beta} + \Psi_{\alpha}(\Psi^{*}_{\beta;\rho;\lambda} - \Psi^{*}_{\beta;\lambda;\rho})]$$
$$+ ([q_{\mu;\rho;\lambda}] - [q_{\mu;\lambda;\rho}]) \tag{3.6}$$

where $\alpha, \beta = 1, 2$ denote the two-component spinor indices.

The last term in (3.6) defines the behavior of q_μ as a four-vector in the curved configuration space. Thus, by definition of the Riemann curvature tensor,

$$[q_{\mu;\rho;\lambda}] - [q_{\mu;\lambda;\rho}] = R_{\kappa\mu\rho\lambda}q^\kappa \tag{3.7}$$

Spin curvature

The 'spin curvature tensor' $K_{\rho\lambda}$ is defined as follows in terms of the 'four-dimensional curl' of a spinor field Ψ:

$$\Psi_{;\rho;\lambda} - \Psi_{;\lambda;\rho} = K_{\rho\lambda}\Psi = (\partial_\lambda\Omega_\rho - \partial_\rho\Omega_\lambda + \Omega_\lambda\Omega_\rho - \Omega_\rho\Omega_\lambda)\Psi \tag{3.8}$$

We see here that the spin curvature tensor is antisymmetric, $K_{\rho\lambda} = -K_{\lambda\rho}$.

Substituting (3.7) and (3.8) into (3.6), we have the following relation:

$$K_{\rho\lambda}q_\mu + q_\mu K_{\rho\lambda}^+ = -R_{\kappa\mu\rho\lambda}q^\kappa \tag{3.9}$$

where the '+' superscript denotes the hermitian conjugate field.

In a similar fashion, the vanishing of the 'four-dimensional curl' of the conjugated quaternion yields the relation:

$$K_{\rho\lambda}^+ q_\mu^* + q_\mu^* K_{\rho\lambda} = R_{\kappa\mu\rho\lambda}q^{\kappa*} \tag{3.10}$$

Multiplying (3.9) on the right with q_γ^* and (3.10) on the left with q_γ, adding and subtracting the resulting equations and using the orthogonality relation:

$$q_\gamma q^{\kappa*} + q^\kappa q_\gamma^* = -2\sigma_0\delta_\gamma^\kappa \tag{3.11}$$

we arrive at the correspondence with the Riemannian curvature tensor as follows:

$$R_{\kappa\mu\rho\lambda} = (1/2)(K_{\rho\lambda}q_\mu q_\kappa^* - q_\kappa q_\mu^* K_{\rho\lambda} + q_\mu K_{\rho\lambda}^+ q_\kappa^* - q_\kappa K_{\rho\lambda}^+ q_\mu^*) \tag{3.12}$$

The Ricci tensor is, by definition, the contraction:

$$R_{\kappa\rho} = R_{\kappa\rho\lambda}^\lambda = g^{\mu\lambda}R_{\mu\kappa\rho\lambda} = (1/2)(K_{\rho\lambda}q^\lambda q_\kappa^* - q_\kappa q^{\lambda*}K_{\rho\lambda}$$
$$+ q^\lambda K_{\rho\lambda}^+ q_\kappa^* - q_\kappa K_{\rho\lambda}^+ q^{\lambda*}) \tag{3.13}$$

The Riemann scalar curvature is then:

$$R = g^{\kappa\lambda}g^{\rho\lambda}R_{\kappa\rho} = (1/2)(K_{\rho\lambda}q^\lambda q^{\rho*} - q^\rho q^{\lambda*}K_{\rho\lambda} + q^\lambda K_{\rho\lambda}^+ q^{\rho*} - q^\rho K_{\rho\lambda}^+ q^{\lambda*})$$
(3.14)

The Quaternion Metrical Field Equations

The Lagrangian density is $L = L_E + L_M$, where L_E yields the metrical field equation and L_M leads to the matter field source on the right-hand side of these field equations. We then take:

$$L_E = \text{Tr}R\sqrt{-g} = (1/2)\text{Tr}(q^{\rho*}K_{\rho\lambda}q^\lambda + \text{h.c.})\sqrt{-g}$$
(3.15)

where $-g = -\det g_{\mu\nu}$ is the 'metric density' and 'Tr' is the trace (the sum of diagonal terms).

The variational calculation that yields the quaternion field equations is $\delta S/\delta q^{\lambda*} = 0$, where $S = \int L d^4x$ is the action function. This gives the quaternion field equation:

$$(1/4)(K_{\rho\lambda}q^\lambda + q^\lambda K_{\rho\lambda}^+) + (1/8)Rq_\rho = kT_\rho$$
(3.16a)

The variational equation $\delta S/\delta q^\lambda = 0$ yields the conjugated quaternion equation:

$$(-1/4)(K_{\rho\lambda}^+ q^{\lambda*} + q^{\lambda*}K_{\rho\lambda}) + (1/8)Rq_\rho^* = kT_\rho^*$$
(3.16b)

where $T_\rho = \delta \int L_M d^4x/\delta q^{\rho*}$ is the energy-momentum quaternion source term.[28]

The quaternion metrical field equation (3.16a) (or its conjugate equation (3.16b)) is the factorization of Einstein's tensor field equations (2.8), corresponding to the 16 relations that reflect the matter field of the universe in terms of the geometrical field that defines the spacetime. This is the formal expression of the unified field theory anticipated by Einstein.

A Symmetric Tensor-Antisymmetric Tensor Representation of General Relativity — Gravity and Electromagnetism

If we multiply (3.16a) on the right with q_γ^* and (3.16b) on the left with q_γ, add and subtract the resulting equations, we obtain the following:

$$(1/2)[K_{\rho\lambda}q^\lambda q^*_\gamma - (\pm)q_\gamma q^{\lambda*}K_{\rho\lambda} + q^\lambda K^+_{\rho\lambda}q^*_\gamma - (\pm)q_\gamma K^+_{\rho\lambda}q^{\lambda*}]$$

$$+(1/4)(q_\rho q^*_\gamma \pm q_\gamma q^*_\rho)R = 2(k(+) \text{ or } k'(-))(T_\rho q^*_\gamma \pm q_\gamma T^*_\rho) \quad (3.17\,\pm)$$

where k (or k') depends on the addition (or subtraction) in the equations (3.17).

The Einstein Field Equations from the Symmetric Tensor Part

Comparing the left-hand side of (3.17+) and the correspondences (3.13), (3.14) and (3.2) for Ricci's tensor, $R_{\rho\lambda}$, the scalar Riemann curvature R and the metric tensor, $g_{\rho\lambda}$, we see that the left-hand side of Eq. (3.17+) corresponds precisely with Einstein's tensor $G_{\rho\lambda} = R_{\rho\lambda} - (1/2)g_{\rho\lambda}R$. Thus, the right-hand side of (3.17+) corresponds with the symmetric energy momentum tensor $T_{\rho\gamma}$. Einstein's field equations (2.8) are then in one-to-one correspondence with the symmetric tensor part of the quaternion form of the field equations (3.17+). These are 10 out of the 16 equations of the quaternion factorization of Einstein's formalism that explains gravity.

The Maxwell Field Equations from the Antisymmetric Tensor Part

The remaining 6 equations (3.17−) will now be seen to yield the anti-symmetric, second-rank tensor formalism that leads to Maxwell's equations for electromagnetism.

We proceed by taking the trace of both sides of (3.17−), giving:

$$R^a_{\rho\gamma} + (1/8)\text{Tr}(q_\rho q^*_\gamma - q_\gamma q^*_\rho)R = k'\text{Tr}(T_\rho q^*_\gamma - q_\gamma T^*_\rho),$$

where, by definition, the 'anti-Ricci tensor' is:

$$R^a_{\rho\gamma} = R^{\lambda a}_{\rho\gamma\lambda} = g^{\lambda\alpha}R^a_{\rho\gamma\lambda\alpha} = (1/4)g^{\lambda\alpha}\text{Tr}[K_{\rho\lambda}(q_\alpha q^*_\gamma + q_\gamma q^*_\alpha) + \text{h.c.}]$$

Since the spin curvature is an antisymmetric tensor, $K_{\rho\lambda} = -K_{\lambda\rho}$, it follows (with (3.12)) that:

$$R^a_{\gamma\lambda\mu\rho} = -R^a_{\lambda\gamma\mu\rho} = R^a_{\gamma\lambda\rho\mu}$$

Thus, we see that $R^a_{\rho\gamma}$ is antisymmetric:

$$R^a_{\rho\gamma} = R^{\lambda a}_{\rho\gamma\lambda} = -R^{\lambda a}_{\gamma\rho\lambda} = -g^{\lambda\alpha}R^a_{\gamma\rho\lambda\alpha} = -g^{\alpha\lambda}R^a_{\gamma\rho\lambda\alpha} = -R^{\alpha a}_{\gamma\rho\alpha} = -R^a_{\gamma\rho}$$

It then follows that the left-hand side of (3.17−) is an antisymmetric second-rank tensor. The right-hand side of (3.17−) must then also transform this way.

The eight Maxwell's equations entail four equations with sources and four without sources. To yield the four Maxwell equations with sources, we take the covariant divergence of (3.17−) and multiply both sides of this equation by the constant Q, with the dimension of electrical charge. Thus, we have the four Maxwell field equations with sources:

$$F^{;\rho}_{\rho\gamma} = (4\pi/c)j_\gamma \qquad (3.18)$$

where the antisymmetric electromagnetic field intensity is:

$$F_{\rho\gamma} = Q[(1/4)(K_{\rho\lambda}q^\lambda q^*_\gamma + q_\gamma q^{\lambda*}K_{\rho\lambda} + q^\lambda K^+_{\rho\lambda}q^*_\gamma + q_\gamma K^+_{\rho\lambda}q^{\lambda*})$$
$$+(1/8)(q_\rho q^*_\gamma - q_\gamma q^*_\rho)R] = -F_{\gamma\rho} \qquad (3.19)$$

The current density source in (3.18) is:

$$j_\gamma = (cQk'/4\pi)(T^{;\rho}_\rho q^*_\gamma - q_\gamma T^{;\rho*}_\rho) \qquad (3.20)$$

The remaining four Maxwell equations — those without sources — are:

$$F_{[\rho\gamma;\lambda]} = 0 \qquad (3.21)$$

where the square bracket denotes a cyclic sum. The zero on the right-hand side of (3.21) follows because of the dependence of $F_{\rho\gamma}$ on the antisymmetric tensor spin curvature $K_{\rho\gamma}$, with the vanishing cyclic sum $K_{[\mu\nu;\lambda]} = 0$.

The latter follows because, in configuration space, $K_{\mu\nu} = \Omega_{\mu;\nu} - \Omega_{\nu;\mu}$ is a four-dimensional curl of a vector field.

Equation (3.21) indicates here, there are no magnetic monopoles predicted, in agreement with the empirical facts.[29]

Conclusions

Summing up, we have seen that the quaternion factorization of Einstein's 10 tensor field equations leads to a set of 16 independent field equations. These transform as a four-vector in configuration space wherein each of the four-vector components are quaternion-valued. These 16 equations are then broken up into 10 second-rank *symmetric tensor* equations and 6 second-rank *antisymmetric tensor* equations. The former 10 equations are in one-to-one correspondence with Einstein's original field equations, thereby explaining gravity. The latter 6 antisymmetric tensor equations were shown to yield the form of Maxwell's equations, thereby explaining electromagnetism. The field equation (3.16a) (or 3.16b) is the unified field theory that was sought by Einstein.

The reason for this unification is the removal of the time and space reflection symmetries from the original Einstein tensor equations. This leaves the remaining symmetry as only the continuous transformations in space and time, as required by the principle of relativity. In view of Noether's theorem, a necessary and sufficient condition for the incorporation of the laws of conservation of energy, momentum and angular momentum (in the flat spacetime limit of general relations in the field theory in general relativity) is that the transformations that preserve the forms of the laws of nature be analytic (i.e. non-singular everywhere). Thus, the underlying symmetry group of general relativity theory is a Lie group — it is called the 'Einstein group'. The solutions of the laws of nature according to this underlying group are then regular — continuous and analytic everywhere. The requirement that only regular solutions are allowed in the field theory was called for throughout Einstein's study of the theory of general relativity.

Finally, it was found that the irreducible representations of the Einstein group obey the algebra of quaternions, whose structure entails a second-rank spinor. Thus, the solutions of the laws of nature, according to this theory, are the spinor and quaternion field variables. It is the reason that the factorized version of general relativity, in giving rise to a unified field theory, is in terms of quaternion and spinor field variables.

In the next chapter we will explore the ramifications of the quaternion formulation of general relativity theory in the problem of cosmology. It yields the spiral structure of an oscillating universe cosmology.

An Oscillating, Spiral Universe Cosmology

Introduction

In the 1920s Edwin Hubble discovered that the universe is expanding. That is, each of the galaxies is moving away from a neighboring galaxy in accordance with the Hubble law, $v = Hr$, where v is their relative speed and r is their mutual separation.[7]

The way that Hubble discovered this was to measure the Doppler Effect of the emitted radiation of distant galaxies. If the galactic emitter of radiation G_1 is moving away from a different galaxy, G_2, its spectral lines will be seen to shift toward the longer wavelength end of the spectrum, compared with the spectra of the galaxy G_2. This is a 'cosmological red shift', characterizing the 'expanding universe'. If the universe is instead contracting, the Doppler Effect would be a 'cosmological blue shift' — a shift toward the shorter wavelength end of the spectrum. Hubble observed a cosmological red shift, thus indicating that the universe is expanding.

If the universe is expanding in this way, is it expanding into empty space? The answer is no. This is because the 'universe' is all that there is, *by definition*. There is no *physical space, as a thing-in-itself*. The actual meaning of 'expanding universe' is that the density of matter, as measured by any observer, anywhere, is decreasing with respect to this observer's time measure. The expansion of the universe then signifies that at increasingly earlier times, the matter of the universe became more and more dense. In the limit, then, in a finite time in the past, the matter of the universe was maximally dense and unstable.

At that point in time, there was an explosion — the 'big bang' — when any of the matter of the universe moved away from all other matter.[8] In time, as the expanding universe cooled, stars and galaxies of stars, as well as planets bound to some of them, were formed, leading to our present view of the heavens.

Cosmologists speculate that the initial distribution of the matter of the universe was homogeneous and isotropic. With this speculation and Einstein's field equations (2.8), the Hubble law was derived. 'Homogeneous' means that the matter of the universe is distributed similarly everywhere. 'Isotropic' means that the distribution of matter of the universe is the same as seen from any angle. Both these assumptions are empirically false. This is so according to the modern-day sophisticated instruments, such as the Hubble telescope, or even observations with the naked eye! Galaxies cluster in certain regions of the universe and are absent in others. Further, the galaxies in themselves are not homogeneous and isotropic distributions of stars.

With this scenario for the expansion, the matter of the universe is monotonically becoming less dense as time progresses. Eventually, all the stars will use up their nuclear fuel and the entire universe will evolve toward a homogeneous sea of cosmic dust, becoming increasingly rarefied.

In my view, the scenario of this 'single big-bang cosmology' has a serious theoretical flaw. It is that it entails an absolute point in time, the time of the big bang, when the creation of the material universe was supposed to have happened, *ab initio*. With this view, all local times may be measured with respect to the *absolute time* of the creation of the universe — called 'cosmological time'. On the other hand, the theory of relativity rejects the concept of an absolute time measure. All time measures, according to the theory of relativity, are relative to the frame of reference in which a law of nature is expressed — even the reference frame of the entire universe.

The empirical flaws in the single big-bang scenario, as currently described in cosmology, are those indicated earlier — that it assumes the matter distribution of the universe is homogeneous and isotropic. *This is a false assumption*. Indeed, there is no physical law that requires the matter distribution of the universe (its galaxies, stars, planets, etc.,) must be homogeneous and isotropic.

The oscillating universe cosmology

The question then arises in connection with the big-bang cosmology: how did the matter of the universe get into the state of maximum density and instability in the first place? The only *scientific answer* I see is that *before* the big-bang event, the matter of the universe was imploding (contracting) from a less dense state. The implosion started at a critical time, when the predominant repulsive forces between the matter of the expanding universe, because of its low density, changed to predominant attractive forces, and a contraction of the universe as a whole ensued. At the next critical time, when the matter became sufficiently dense, the contraction changed to an expansion again, where the forces became predominantly repulsive, to continue until the next inflection point when the expansion would once again turn into a contraction, and so on, *ad infinitum*. This is the 'oscillating universe cosmology'.

What is important, theoretically, is that the oscillating universe cosmology is compatible with the requirements of the theory of relativity while the single big-bang cosmology is not. With the oscillating universe, there is no absolute 'beginning'. The last big bang, around 15 billion years ago, was then not a unique occurrence; it was only the beginning of this particular cycle of the oscillating universe.

Equations of motion in general relativity

The equation of motion of an unobstructed body in classical physics is: $x^k(t)'' = 0$, where $k = 1, 2, 3$ are the three spatial coordinates and the $''$ refers to the second derivative with respect to the time t. The solution of this equation of motion, after integration, implies that matter moves on a straight line at a constant speed, so long as it is unobstructed by external means. This is Galileo's *principle of inertia* (also known as 'Newton's first law of motion'). If there would be a force applied to the body, in the kth direction, F^k, the generalization of this equation becomes: $x^k(t)'' = F^k/m$, where m is the inertial mass of the body. This is 'Newton's second law of motion'.

In general relativity, the equation of motion of an unobstructed body, in the curved spacetime, is the geodesic equation, $x^\mu(s)'' = -\Gamma^\mu_{\nu\lambda} x^\nu(s)' x^\lambda(s)'$, where the parameter '$s$' plays the role of the proper time. Thus, the affine connection term on the right-hand side of this equation of motion plays the role of the force that acts on the body. The latter is then the *geometrical reflection* of the potential force that acts on this body as it propagates in the curved spacetime. *This is the essence of the principle of equivalence of general relativity theory.*

Since the affine connection coefficients $\Gamma^\mu_{\nu\lambda}$ are not positive definite functions, the effective force on the body in the curved spacetime could be either repulsive or attractive, depending on physical conditions. If they are repulsive forces, due to a high density of matter and relative speeds of interacting matter that are close to the speed of light, then any matter of the universe would move away from other matter and the universe would be in the expansion phase. When the matter of the universe becomes sufficiently rarefied and the relative speeds are small, compared with the speed of light, there would be an inflection point where the affine connection terms change sign and thus the effective forces become attractive, leading any matter of the universe to move toward other matter. This would be the contracting phase of the universe. In the latter phase, the matter would become increasingly dense until the next inflection point, where the attractive forces would become dominated by repulsive forces, starting the expanding phase of the oscillating universe once again.

With this cosmology, the presently observed expansion of the universe is only a phase of an ever-oscillating universe, alternating between expansion and contraction. The last big bang, which estimates from the Hubble law predict to be about 15 billion years ago, is only the beginning of this particular cycle of the oscillating universe cosmology.

One might ask the question: 'When did all of the cycles of the oscillating universe begin?' That is, when did the actual creation of the universe take place? Thus cannot be answered in the context of science. It is a theological question based on the concept of religious truth. This is a non-refutable sort of truth, based on faith, while scientific truth is refutable, and based on empirical confirmation and logical consistency in its expression. To answer a scientific question with a theological answer would be a non-sequitur because their underlying logics are different. In regard to the reference to G-d as the Creator of the universe, it is interesting that in Jewish philosophy, Kabbalah, G-d is sometimes referred to (in Hebrew) as: haya-hoveh-yeeheeyeh (was-is-will-be). That is, in the context of G-d and His creation, the concept of the progression of 'time' has no meaning.

In the next part of this chapter, we will demonstrate in mathematical terms that the quaternion form of general relativity is compatible with an oscillating universe cosmology with a spiral configuration. The same quaternion formalism also predicts the rotations of the galaxies and their own spiral configurations. In Chapter 5, on the subject of 'dark matter', it will be seen that it is a rotating spiral universe that predicts, in the context of this theory, the separation of matter from antimatter in the 'big bang' periods of the spiral universe at the initial stages of its expansion phases.

Dynamics of the Expansion and Contraction of the Universe

The geodesic equation in quaternion form

As it is shown in many treatises and texts on general relativity theory, the geodesic equation follows from the variational minimization of path length in the curved spacetime, $\delta \int ds = 0$.[30] Taking ds to be the quaternion form $ds = q^\mu dx_\mu$, as shown earlier, we arrive at the quaternion-valued geodesic equation[31]:

$$[x^{\mu\prime\prime}(s) + \Gamma^\mu_{\nu\lambda} x^{\nu\prime}(s) x^{\lambda\prime}(s) = 0]_{\alpha\beta} \qquad (4.1)$$

where α, β = (1,1), (2,2), (1,2), (2,1) denote the four equations that predict the path of an unobstructed body $x(s)$ and the primes denote differentiation with respect to the interval 's'. These four equations must be solved simultaneously in order to predict the trajectory of a body. This is a generalization of the real number-valued time parameter in the standard theory to a quaternion number-valued time parameter. That is, with this formalism, to follow the body along its geodesic path, one needs four parameters at each spatial point to proceed to the next continuously connected spatial point of its path. Here, the parameter 's' plays the role of the proper time for the trajectory. *This quaternion generalization of the time parameter in the laws of physics was anticipated by William Hamilton with his discovery of the quaternion algebra, in the 19th century.*

The solution of the geodesic equation $x(s)$ is the path of the body in the curved spacetime, as we discussed in the preceding section. *It is a generalization of Galileo's principle of inertia.*

Let us now parameterize the variation with respect to the quaternionic line element. In the conventional real number tensor theory, one has the differential operator:

$$d/ds = \lim_{\Delta x^\mu \to 0, \, \Delta x^\nu \to 0}[\Delta/\Delta(g_{\mu\nu}x^\mu x^\nu)^{1/2}] = d/(g_{\mu\nu}dx^\mu dx^\nu)^{1/2},$$

where, by definition of the derivatives, the metric tensor $g_{\mu\nu}$ is evaluated at the spacetime point where the derivatives are taken.

In the quaternion calculus, we must instead utilize the differential operator:

$$d/ds = \lim_{\Delta x^\mu \to 0}[\Delta/\Delta(q_\mu x^\mu)] = q_\mu^{-1}d/dx^\mu \qquad (4.2)$$

in which q_μ^{-1} is the inverse of q_μ, evaluated at the spacetime point where the derivatives are taken. According to the algebra of quaternions, the inverse of the quaternion is as follows:

$$q_\mu^{-1} = q_\mu^*/q_\mu q_\mu^* \qquad (4.3)$$

The second derivative of the quaternion variable is then:

$$d^2/ds^2 = q_\mu^{-1}d/dx^\mu[q_\mu^{-1}d/dx^\mu] \qquad (4.4)$$

The vanishing variation $\delta \int ds = \delta \int q_\mu dx^\mu = 0$ then gives rise, in the same manner as the derivation of the conventional geodesic equation, to the same functional form of this equation, shown in (4.1).

The particular problem of determining the path of a test body need not entail the noncommutative feature of quaternions, e.g. when polarization does not play a role in the application at hand. With this in mind, a version of the geodesic equation (4.1) is in terms of the determinant of this equation,

$$d^2 x^\mu / dS^2 + \Gamma^\mu_{\nu\lambda} dx^\nu / dS \, dx^\lambda / dS = 0 \tag{4.5}$$

where

$$dS = |q_\mu| dx^\mu, \quad d/dS = |q_\mu^{-1} d/dx^\mu| = |q_\mu|^{-1} d/dx^\mu \tag{4.6}$$

and the vertical bars denote the determinant with respect to the spinor indices. Equation (4.5) is the determinant of the geodesic equation (4.1).

The real number form (4.5) of the geodesic equation has the same functional form as the standard form of the geodesic equation. Nevertheless, they are different because the latter is in terms of derivatives with respect to the differential $ds = (g_{\mu\nu} dx^\mu dx^\nu)^{1/2}$ while the former is with respect to $dS = |q_\mu| dx^\mu$. The form of the geodesic equation (4.5) applied to stationary state problems (such as orbital motion) is not the same as the conventional one. This is because of the time-dependence that is implicit in q_μ, as compared with the time-independence of $g_{\mu\nu}$ in the stationary state problem.

Let us now commence by defining the interval dS in the reference frame wherein the test body is located at the spatial origin. In this reference frame,

$$d/dS = |q_0|^{-1} d/dX^0,$$
$$d^2/dS^2 = |q_0|^{-2} d^2/dX^{02} + |q_0|^{-1} d/dX^0 |q_0|^{-1} d/dX^0 \tag{4.7}$$

where X^0 is the time measure in the reference frame that is in motion with respect to the global coordinate frame.

It follows from (4.7) that the geodesic equation (4.5) may be expressed in the form:

$$|q_0|^{-2}x^{\mu''} + |q_0|^{-1}(|q_0|')^{-1}x^{\mu'} + \Gamma^{\mu}_{\nu\lambda}x^{\nu'}x^{\lambda'} = 0 \qquad (4.8)$$

This is to be taken as the equation of motion of a test body. It is equivalent to the form:

$$x^{\mu''} + \Gamma^{\mu}_{\nu\lambda}x^{\nu'}x^{\lambda'} = -(1/2)Q^{-1}Q'x^{\mu'} \qquad (4.9)$$

where $Q = |q_0|^{-2}$.

When $v/c \ll 1$, it follows that to order v/c, the time derivatives of the global coordinates are $x^{0'} = c$, and $x^{0''} = 0$. With $\mu = 0$, Eq. (4.9) then becomes:

$$\Gamma^{0}_{\nu\lambda}x^{\nu'}x^{\lambda'} = -(1/2)cQ^{-1}Q' \qquad (4.10)$$

The equation of motion in terms of the spatial coordinates ($k = 1, 2, 3$) is then:

$$x^{k''} + \Gamma^{k}_{\nu\lambda}x^{\nu'}x^{\lambda'} = \Gamma^{0}_{\nu\lambda}x^{\nu'}x^{\lambda'}(x^{k'}/c) \qquad (4.11)$$

where the ' and " refer to first and second order differentiation with respect to the independent time parameter t.

Thus, the explicit generalization of the geodesic equation (to order v/c) that follows from the quaternionic expression for the stationary state in general relativity entails a non-zero term on the right-hand side of (4.11) — a term that is non-invariant under time reversal. This implies a damping feature of the oscillating behavior of the test body, such as the case of planetary motion.

Dynamics of the Oscillating Universe Cosmology

With the assumption that the terms $\Gamma^{k}_{\mu\nu}x^{\mu'}x^{\nu'}$ in the equation of motion (4.11) are time-independent over the time scale of observation, this equation may be integrated, yielding the solution:

$$x^k(t) = K_1 + K_2 \exp{(c^{-1}\Gamma^{0}_{\mu\nu}x^{\mu'}x^{\nu'}t)} + (\Gamma^{k}_{\mu\nu}x^{\mu'}x^{\nu'}/\Gamma^{0}_{\mu\nu}x^{\mu'}x^{\nu'})ct \qquad (4.12)$$

where K_1 and K_2 are the two integration constants of the second order differential equation (4.11), to be determined by the boundary conditions.

Assuming now the oscillating universe cosmology as the only covariant model, there is an inflection point in the velocity of a test body, dx^k/dt, at the times of alternation between the expansion and contraction phases. Calling $t = 0$ the time when the presently observed expansion phase occurred in our cycle of the oscillating universe, and locating the test matter at the origin, the boundary condition imposed by the cosmology are:

$$dx^k/dt(0) = x^k(0) = 0 \tag{4.13}$$

With (4.13) in (4.12), the integration constants are found to be:

$$K_1 = -K_2 = c^2[\Gamma_{\mu\nu}^k x^{\mu'} x^{\nu'} / (\Gamma_{\mu\nu}^0 x^{\mu'} x^{\nu'})^2]_0 \tag{4.14}$$

where the '0' subscript of the square bracket refers to the value at the beginning of a cycle of the oscillating universe.

With the assumption that $dx^k/dt \ll c$, Eq. (4.14) becomes:

$$K_1 = -K_2 = [\Gamma_{00}^k / (\Gamma_{00}^0)^2]_0$$

and the solution (4.11) takes the form:

$$x^k(t) = [\Gamma_{00}^k / (\Gamma_{00}^0)^2]_0 [1 - \exp(c^{-1}\Gamma_{\mu\nu}^0 x^{\mu'} x^{\nu'} t)]$$
$$+ (\Gamma_{\mu\nu}^k x^{\mu'} x^{\nu'} / \Gamma_{\mu\nu}^0 x^{\mu'} x^{\nu'})ct \tag{4.15}$$

The coefficient that multiplies the first term on the right side of (4.15) has the dimension of length:

$$[\Gamma_{00}^k / (\Gamma_{00}^0)^2]_0 = R^k \tag{4.16}$$

This may be interpreted as the radius of the universe when applied to the furthermost stellar objects of the night sky. When R^k is large and spacelike compared with ct, the second term on the right-hand side of (4.15) may be neglected, compared with the first. This approximation corresponds to the assumption that a distance from the observer to a faraway galaxy is large, compared with the distance traveled by light from one time that the astronomer observes the galaxy to the next.

Derivation of the Hubble Law as an Approximation

With the latter assumption, the resulting solution predicts that the speed of a galaxy in any of the spatial directions x^k is:

$$dx^k/dt = v^k = c^{-1}\Gamma^0_{\mu\nu}x^{\mu'}x^{\nu'}(x^k - R^k) \qquad (4.17)$$

This is an expression of the Hubble law. Since R^k cancels in the comparison of two speeds at the corresponding times, (4.17), then predicts the speed of a galaxy relative to another galaxy, v^k, is linearly proportional to their separation, x^k. This assumes the term in front of the right-hand side of (4.17) is constant in time. The latter term is then the 'Hubble constant',

$$H = c^{-1}\Gamma^0_{\mu\nu}x^{\mu'}x^{\nu'} = -(1/2)Q^{-1}Q' \qquad (4.18)$$

The quaternion factor, $Q = |q_0|^{-2}$ was previously defined in eq. (4.9) in terms of the determinant of the time component of the quaternion field q_0.

With the definition (4.18) of the 'Hubble constant', H, we see that it is actually dependent on the time parameter. But with the approximation used, it appears to be time-independent over the time span that is observed in our view of the expanding universe. Nevertheless, independent of approximations, H must be space- and time-dependent. Thus, the Hubble law (4.17) is a nonrelativistic approximation to describe the expanding universe because of the covariance of the laws of nature, including the physical dynamics of cosmology.

The Spiral Structure of the Universe[32]

With the assumption of the boundary conditions (4.13) applied to the constituent matter of the universe, subject to the rest of the matter of the universe, we will assume that the radius of the system R^k is the same order of magnitude as ct. Thus, we may not, in this case, neglect the second term on the right-hand side of (4.15).

In this case, however, we may consider the motion of the test mass over time intervals that are short compared with the time it would take light to traverse the entire domain of the universe. Under these conditions, and assuming the speed of the test matter relative to the 'center of mass' of the entire universe is small, compared to the speed of light c, the argument of the exponential factor in (4.15) is small compared with unity, so that,

$$c^{-1}\Gamma^0_{\mu\nu}x^{\mu'}x^{\nu'}t \approx \Gamma^0_{00}ct \approx ct/R \ll 1$$

where R is the order of magnitude of the interaction domain radius of the entire universe, according to Eq. (4.16).

With this approximation, the expansion of the exponential in (4.15), keeping the first two terms, gives the location of the moving matter as follows:

$$x^k(t) = ct[\Gamma^k_{00}/(\Gamma^0_{00})^2 - [\Gamma^k_{00}/(\Gamma^0_{00})^2]_0]\Gamma^0_{00} \qquad (4.19)$$

With the expression for the affine connection in terms of the metric tensor,

$$\Gamma^\rho_{\mu\nu} = (1/2)g^{\rho\lambda}(\partial_\mu g_{\lambda\nu} + \partial_\nu g_{\mu\lambda} - \partial_\lambda g_{\nu\mu}) \qquad (4.20)$$

and assuming a Taylor expansion of the metric tensor, as an analytic function of the time parameter t, it follows that for small t the spatial coordinates of the test mass are given by the relation:

$$x^k(t) = (1/2)a^k t^2 + b^k t^3$$

where a^k and b^k are functions of the first and second derivatives of $g_{\mu\nu}(t)$.

With this result, the acceleration of the test mass in the kth direction has the form:

$$d^2 x^k/dt^2 = a^k + 6b^k t \qquad (4.21)$$

Substituting the variables $d^2\varsigma^k/dt^2 = d^2x^k/dt^2 - a^k$, (4.21) leads to the form:

$$(d^2\varsigma^1/dt^2 + d^2\varsigma^2/dt^2 + d^2\varsigma^3/dt^2) = 36[(b^1)^2 + (b^2)^2 + (b^3)^2]t^2$$

$$(4.22)$$

The metrical field q_μ is a spin-one variable, implying that in addition to translational motion, there is rotational motion of the test body that moves on the quaternionic geodesic path. Thus, the entire closed system (in principle, the universe) must be set into rotational motion in a plane that is perpendicular to the orientation of the imposed spin of the system as a whole. Thus, if x^3 is (defined to be) the direction of the axis of rotation of the closed system, the quaternion field equations must predict that $(b^1, b^2) \gg b^3$. We may then formulate a (non-isotropic) two-dimensional displacement of the test mass (a constituent of the universe near the beginning of one of its expansion phases) in terms of the vector:

$$\varsigma^r = e_x\varsigma^x + e_y\varsigma^y$$

This is in reference to an arbitrarily chosen origin at $x = 0$, $y = 0$ that defines the initial motion of a test body. The nonlinear differential equation in terms of this coordinate system is then:

$$(d^2\varsigma^r/dt^2)^2 = (d^2\varsigma^x/dt^2)^2 + (d^2\varsigma^y/dt^2)^2 = A^2t^2 \qquad (4.23)$$

where $A^2 = 36[(b^x)^2 + (b^y)^2]$.

Clearly, the coordinate of the test body must obey the boundary conditions of the oscillating cosmology, $\varsigma^r(0) = 0, d\varsigma^r/dt(0) = 0$.

With these boundary conditions, the solutions of (4.23) are the *Fresnel integrals*:

$$\varsigma^x(t) = c\left\{\int_0^t \cos[(A/2c)\tau^2]d\tau - t\right\}, \quad \varsigma^y(t) = c\int_0^t \sin[(A/2c)\tau^2]d\tau$$

$$(4.24)$$

This solution defines the *Cornu Spiral*.

The total solution for the test body in the universe in this 'early time' of an expansion phase (just after a 'big bang') is then:

$$x(t) = \varsigma^x(t) + (1/2)a^x t^2, \quad y(t) = \varsigma^y(t) + (1/2)a^y t^2 \qquad (4.25)$$

It is then readily verified with this solution that $x(0) = dx/dt(0) = 0$, in accordance with the boundary conditions of the oscillating universe cosmology.

The solution (4.25) is a superposition of the spiral motion of a test body of the universe in a two-dimensional plane, characterized by the Fresnel integrals (4.24), and a constant acceleration relative to an observer's frame of reference. In the rest frame of the test body itself, $a = 0$. In this reference frame, one then has a purely spiral motion with two inflection points — one at the beginning of an expansion phase, when there is a maximum density of matter and the dominant gravitational force is repulsive, and the other at minimum matter density, when the dominant gravitational force is attractive, leading to the beginning of the contraction phase of the cycle. These oscillations of the matter of the universe then continue indefinitely, into the past and into the future.

Concluding Remarks

The cosmological model discussed in this chapter is then in contrast with the 'single big bang' model of present-day consensus, wherein there is a singular beginning of all matter in space and time, and then an explosion and a single, unique expansion. With the single big bang cosmological model, the distribution of matter of the universe, at the 'beginning', was isotropic and homogeneous, and continues to be so. The observations of the galaxies and their clustering are convincing that this model is not true to nature. Nor is its assumption of an absolute cosmological time, initiating at the time of the assumed *singular* big bang. Theoretically, these ideas do not conform with the covariance requirement of the theory of relativity, as a basis for cosmology, as the oscillating universe cosmology does.

A further remark is that the same boundary conditions and approximations for the oscillating universe cosmology could also apply to the individual galaxies, such as our own Milky Way. The same predictions must then follow, that a constituent star of a galaxy evolves along a spiral path in the confines of the mother galaxy. It would have two inflection points in its motion, yielding an oscillating galaxy in itself, continually expanding and contracting — until the stars would eventually burn themselves out. The empirical observation of the spiral structures of most of the galaxies attests to this theoretical astrophysical prediction.

In the next chapter we will discuss a model for the dark matter that is assumed to be embedded in the universe.

Dark Matter

Introduction

In the preceding decades, astrophysicists have studied the rotations of the galaxies. It was believed initially that the driving force causing their rotations is the gravitational pull on them by their neighboring galaxies, such as the gravitational force of the galaxy Andromeda on our own galaxy, Milky Way. However, with the knowledge of the approximate masses of the neighboring galaxies, and using Newtonian dynamics as a first approximation for their mutual forces, it was found that this was insufficient to cause the observed rotational dynamics of the galaxies. It was then speculated that there must be some invisible matter permeating the universe that is gravitationally coupled to the galaxies, that causes their rotations. This was called 'dark matter'.

It will be proposed in this chapter that the dark matter is a dense sea of particle-antiparticle pairs (electron-positron and proton-antiproton), each in their (derived) ground states of null energy (and null linear and angular momentum). The state of null energy corresponds to the 'zero energy' of the bound state of the pair, relative to the state when the particle and antiparticle are free of each other at their combined energy, $2\,mc^2$, where m is the inertial mass of each of these particles. These features will be described in this chapter, in terms of the important steps that lead to them. *The full details of these derivations will be referred to in another publication.*[33] The pairs are electrically neutral but they have gravitational manifestations associated with their masses.

With this result for the 'true' ground state of the particle-antiparticle pair, it is argued that there is no reason for there not to be a dense sea of such pairs in the universe. This has led to the prediction of the masses of the elementary particles as well as the results of blackbody radiation.[34] In the latter, instead of a sea of photons in the cavity at a fixed temperature, the cavity is filled with a sea of pairs in their ground states. This yields the same blackbody radiation curve as the model of a sea of photons, from a statistical analysis, yet without the need of photons to explain the phenomenon.

As far as the masses of elementary particles are concerned, this is shown, explicitly, to be a function of the rest of the matter of a closed system (in accordance with the Mach principle) expressed in terms of the spin-affine connection.[35] A prediction then follows that for every spin one-half matter field, there is a ground state mass and a heavier mass. That is, general relativity theory predicts that the spin one-half particles must occur in *mass doublets*. The scenario in reaching the heavy mass state of a mass doublet is as follows: if an ordinary spin one-half particle, such as an electron, comes close to a pair of the background sea of pairs, it can excite the pair out of its ground state, to a higher energy state. This, in turn, would change the spin-affine connection in the vicinity of the given electron, which in turn would alter the mass of the particle to its higher mass value.

It was found, in a first approximation, that the ratio of the mass of the 'heavy electron' of the electron mass doublet, to that of the lighter electron is $3/2\alpha \approx 206$, where $\alpha \approx 1/137$ is the fine structure constant.[36] It was found, further, that the lifetime of the excited pair, that gave rise to the higher electron mass, is the order of 10^{-6} seconds.[37] These numbers correspond, empirically, to the mass of the muon and its lifetime, as it decays to $e + \nu + \nu^*$. It was found in this analysis that the neutrino ν and its antineutrino ν^* are represented by the spinor electromagnetic fields for the pair, φ_α and φ_α^+, solving the Weyl equation (and its conjugate equation) for the neutrino field. The electromagnetic energy in the latter fields of the decay of the heavy electron is then transferred to the sea of pairs of the surroundings. *Thus, general relativity explains the existence of the muon in nature.* Further, if more than one pair of the background sea of pairs is excited, other higher mass values of the electron ('leptons') would be predicted, such as the tau meson, and even higher mass values as more pairs in the vicinity of the electron are excited. Thus, this theory predicts an infinite spectrum of 'leptons'; rather than the three leptons — electron, muon and tau — of the standard model of contemporary elementary particle physics.

In the next section, we will show the explicit ground state of the pair and that it corresponds to null energy and null linear momentum.

The Field Equations and the Ground State Solution for the Bound Particle-Antiparticle Pair

We start by considering an electron-positron (or proton-antiproton) bound pair. In the special relativity limit of the theory, with the Dirac formalism (reflection symmetry plays no role here), the wave equations for the electron and the positron have the form:

$$(\gamma^\mu \partial_\mu - I(e^+) + \lambda)\psi^{(e^-)} = 0 \qquad (5.1a)$$

$$(\gamma^\mu \partial_\mu - I(e^-) + \lambda)\psi^{(e^+)} = 0 \qquad (5.1b)$$

where λ is the mass of the particle or antiparticle, and γ^μ are the four Dirac matrices.[38] $I(e^\pm)$ are the interaction functionals representing the electromagnetic action of e^+ (or e^-) on e^- (or e^+). Since $I(e^+)$ depends on the matter field wave function $\psi^{(e^+)}$ and the equation in the latter function depends on $I(e^-)$, which in turn depends on $\psi^{(e^-)}$, the operator $I(e^+)$ in the latter equation (5.1a) depends on the solution $\psi^{(e^-)}$ itself. *Thus the wave equations (5.1ab) are intrinsically nonlinear.*

It is found in a complete analysis that an *exact bound state* solution of (5.1ab) is the four-component Dirac spinor[33]:

$$\psi^{(e^+)} = -\psi^{(e^-)} = |\exp(-\lambda t) \quad 0 \quad 0 \quad \exp(\lambda t)\rangle \qquad (5.2)$$

The same analysis shows that:

$$I(e)\psi^{(e)} = 0 \qquad (5.3)$$

It is important to note that while the operation of the functional $I(e)$ on the bound state function $\psi^{(e)}$ yields a zero value, the function I itself is not a null operator. The result (5.3) is a consequence of the nonlinear features of the coupled matter equations (5.1) for the pair — under the special conditions specified in this analysis.[33] It should also be noted that the solution (5.2) is expressed in the proper frame of the pair. In any other Lorentz frame the argument of the exponential would generalize to $k^\mu x_\mu = \mathbf{k} \cdot \mathbf{r} - \lambda t$, where the wave vector $\mathbf{k}$ takes account of the motion of the pair, relative to an observer.

In the preceding analysis, the spinor form of the electromagnetic field equations was used, as its most general expression, since its covariance is only with respect to the continuous transformations in time and space. This is the two-component spinor form of electromagnetic field theory:

$$\sigma^\mu \partial_\mu \varphi_\alpha = \Upsilon_\alpha \tag{5.4}$$

where $\sigma^\mu \partial_\mu = \sigma^0 \partial_0 - \sigma^k \partial_k$ is the first order quaternion differential operator, whose basis elements are σ^0, the unit two-dimensional matrix, and (with the summation convention), $\sigma^k (k = 1, 2, 3)$ are the Pauli matrices.

The correspondence with the magnetic and electric variables (H_k, E_k) of Maxwell's equations is as follows:

$$G_k = H_k + iE_k \quad (k = 1, 2, 3)$$

$$\varphi_1 = |G_3 \quad G_1 + iG_2\rangle, \quad \varphi_2 = |G_1 - iG_2 \quad - G_3\rangle \tag{5.5a}$$

$$\Upsilon_1 = -4\pi i |\varrho + j_3 \quad j_1 + ij_2\rangle = e\psi^+ \gamma^0 \Gamma_1 \psi \tag{5.5b}$$

$$\Upsilon_2 = -4\pi i |j_1 - ij_2 \quad \varrho - j_3\rangle = e\psi^+ \gamma^0 \Gamma_2 \psi \tag{5.5c}$$

$$j_\mu = (\varrho; j) = ie\psi^+ \gamma^0 \gamma_\mu \psi \tag{5.5d}$$

Thus, Eq. (5.4) has the following form for the particle and anti-particle:

$$\sigma^\mu \partial_\mu \varphi_\alpha^{(e^-)} = -e\psi^{(e^-)} + \gamma^0 \Gamma_\alpha \psi^{(e^-)} \tag{5.6}$$

$$\sigma^\mu \partial_\mu \varphi_\alpha^{(e^+)} = e\psi^{(e^+)} + \gamma^0 \Gamma_\alpha \psi^{(e^+)} \tag{5.7}$$

where $e^+ = -e^- = e$ is the electrical charge of the particle and antiparticle. This analysis applies to the electron-positron pair as well as the proton-antiproton pair.

It follows from Noether's theorem that the covariance of the field equations with respect to continuous transformations of the space and time coordinates leads respectively to the conserved linear momentum P_k of the bound pair, as follows:

$$P_k = \int \sum_{i=1}^{n} (\partial L / \partial (\partial_0 \Lambda_\varsigma^{(i)})) \partial_k \Lambda_\varsigma^{(i)} d\mathbf{r} \tag{5.8}$$

with $k = 1,2,3$ being the three spatial directions. The conserved energy is:

$$P_0 = \int \sum_{i=1}^{n} [(\partial L / \partial (\partial_0 \Lambda_\varsigma^{(i)})) \partial_0 \Lambda_\varsigma^{(i)} - L] d\mathbf{r} \tag{5.9}$$

where L is the total Lagrangian density for the pair. It is the sum of two parts, $L_D + L_M$. The former gives the matter behavior of the pair and the latter gives the spinor form of the electromagnetic fields, as in Eq. (5.4). They are as follows:

$$L_D = (hc/2\pi) \sum_{u=1}^{2} \{\psi^{(u)+} \gamma^0 (\gamma^\mu \partial_\mu + I(u) + \lambda(e)) \psi^{(u)} + \text{h.c.}\} \tag{5.10}$$

$$L_M = ig_M \sum_{u \neq v=1}^{2} \sum_{\alpha=1}^{2} (-1)^\alpha \varphi_\alpha^{(u)+} (\sigma^\mu \partial_\mu \varphi_\alpha^{(v)}$$

$$- 2e^{(v)} \psi^{(v)+} \gamma^0 \Gamma_\alpha \psi^{(v)}) + \text{h.c.} \tag{5.11}$$

where (u, v) denotes the bound particle and antiparticle components and 'h.c.' denotes the hermitian conjugate of the preceding function.

The variation of L_M with respect to the spinor field variables φ_α gives the spinor version of the Maxwell field equations (5.4). The variation of L_D with respect to the matter field (bispinor) variables ψ gives the matter field equations (5.1).

The summations in (5.8) and (5.9) are over the field variables $\{\Lambda_\varsigma^{(i)}\}$, where ς denotes the field components. In the case of the bound particle-antiparticle, there are twelve such fields:

$$\{\psi^{(e^-)+}\gamma^0, \psi^{(e^+)+}\gamma^0, \psi^{(e^-)}, \psi^{(e^+)}, \varphi_\alpha^{(e^-)+}, \varphi_\alpha^{(e^+)+}, \varphi_\alpha^{(e^-)}, \varphi_\alpha^{(e^+)}\} \quad (\alpha = 1, 2)$$

With the fields for the pair indicated above, the matter field solutions for the pair (5.2), and the following spinor solutions of (5.4) for the electromagnetic field,[39]

$$\varphi_1 = (4\pi e/\lambda)|1 \quad \exp(2i\lambda t)\rangle \quad \varphi_2 = (4\pi e/\lambda)|\exp(-2i\lambda t) \quad 1\rangle \quad (5.12)$$

it is found[33] that the conserved energy and momentum of the pair in this bound (ground) state is:

$$P_0 = P_1 = P_2 = P_3 = 0 \tag{5.13}$$

That is, the conserved energy-momentum of the pair in this state is a null vector, with each component equal to zero. Thus, if all four components of P_μ are separately zero in one Lorentz frame, they must be zero in any other Lorentz frame. The result derived for the ground state of the pair — the state of minimum energy and momentum — is then Lorentz invariant. *This is the derived energy-momentum of the 'vacuum state', though it is not a vacuum — it is a dense sea of particle-antiparticle pairs, each in its ground state of null energy, momentum and angular momentum, filling the universe. This sea of pairs is the candidate proposed here for the dark matter of the universe.*

Separation of Matter and Antimatter in the Universe

The scenario for the separation of matter from antimatter in the universe is as follows: at the initial stages of the expansion phase of each of the cycles of the oscillating universe, (the 'big bang' stages), the matter of the universe, embedded in a dense sea of particle-antiparticle pairs, begins its expansion phase in a spiraling, rotational motion (as we discussed in the preceding chapter). With the enormous energy present, some of it is delivered to a small fraction of the background pairs, to dissociate them. The positively and negatively dissociated particles then continue in the spiral motion caused by gravitation. The rotation of the positively charged dissociated matter then gives rise to a magnetic field in one direction, parallel to the axis of rotation of the spiraling matter of the universe. The rotation of the negatively charged matter gives rise to a magnetic field in the opposite direction.

The oppositely polarized magnetic fields so-formed then competes with the gravitational field with a single rotational orientation. Thus, many of the positively charged particles (positrons and protons) are then swept by the magnetic field in one direction in the universe (the 'cyclotron effect') while the negatively charged matter (electrons and antiprotons) are swept in the opposite direction in the universe. In this way, as the universe continues to expand and cool, matter and antimatter are sent to distantly remote parts of the universe. The electrons e^-, with higher mobility than the antiprotons p^- (because of their much smaller inertial mass) then bond with positively charged matter of the underlying pairs more abundantly than the antiprotons p^- accompanying them. Similarly, the positrons, e^+, in a different part of the universe, with a higher mobility than the accompanying protons p^+, bind more abundantly with the negatively charged matter of the pairs of the background (i.e. the 'dark matter'). In this way, there results a domain of the universe that is predominantly matter (as is our region of the universe) and different domains that are predominantly antimatter.

Summing up, the separation of matter and antimatter results from a competition between the gravitational fields in the initial formation of the spirally rotating expanding phase of the cycle of an oscillating universe and the magnetic fields that are created by the dissociation of pairs into matter and antimatter. Each of these components moves in opposite directions in these magnetic fields, compared with the single direction of the gravitational field of the spiraling universe.

Olber's Paradox

Another one of the older puzzles of astrophysics is the following: with the indefinitely large number of stars of the universe continually emitting light (and other radiation) since the very beginning of its expanding phase (the order of 15 billion years ago), why is the night sky dark? That is to say, if the accumulation of 'photons' in the universe has been increasing, since the last big bang, to a relatively infinite number, why is it that the night sky is not explosively bright to any observer, rather than being dark? This question is called 'Olber's paradox'.

One of the commonly-held answers to this paradox is that the 'photons' released from the stars of an expanding universe decrease their measured frequencies because of the Doppler Effect. This effect predicts that the frequency v of the radiation from an emitter that recedes from the observer of this radiation, decreases. In the limit, the frequency of this radiation goes to zero. Since the energy of the photon (of the light from the stars) depends linearly on its frequency in accordance with the quantum condition, $E = hv$, it is claimed that the *vanishing* of the measured frequency of radiation from very distant stars, for the observer, and hence its vanishing energy, answers Olber's paradox.

This explanation is not satisfactory to this author since what we are concerned with in the problem of the (*objective*) energy content of the radiation of the universe is the 'proper frequency', not the (*subjectively*) observed frequency due to the Doppler Effect. The 'proper frequency' is related to the *intrinsic* energy of the radiation, independent of the measurements of an observer!

I believe that the answer to the paradox is that there are no 'free photons' in the first place. Light, *per se*, in this view, is only concerned with the emitter of the radiation and the absorber of this radiation, without the idea of a free photon that moves about on its own in the universe. Indeed, this was the view of the quantum of radiation, as related only to the emitter and absorber of this radiation, according to Max Planck, the discoverer of the quantum of light. It was also an idea of Michael Faraday, in the 19th century, in his analysis of the propagation of light.

In this view, matter interacts with matter according to the idea of '*delayed action at a distance*'. A source of radiation emits light only if, at the later time when the light reaches the absorber, it will be there to absorb it![40] According to the theory of relativity, an interaction propagates at the speed of light c between an emitter and an absorber, in the time t, provided that they are at a *timelike* separation $R < ct$. But there are no 'free photons' that fill the universe, as Olber's paradox presumes! On the other hand, if the emitter is at a *spacelike* separation from the absorber (the observer) $R > ct$, there will be no emission from the emitter to the absorber, i.e. there is no interaction between them. Most of the stars of the night sky are at a *spacelike* separation from us. Thus, the night sky is dark to us. This answers Olber's paradox. It is an answer not connected with the existence of dark matter in the universe. Rather, it is explained with the theory of relativity and our definition of 'light', as seen by Planck, as *an electromagnetic interaction* that propagates between an emitter and an absorber, but not as a 'thing in itself'. This view then eliminates the 'photon' as an elementary particle. It is rather something, here, that stands for the interaction coupling between material components, emitter and absorber. (*It is shown in this author's research program that all experiments that are supposed to entail 'photons', as elementary particles, are explained instead in terms of electrically charged matter alone. This is in line with the theory of 'delayed-action-at-a-distance' mentioned above. That is to say, in this view the 'photon' is a superfluous entity in physics.*[16,17])

6

Concluding Remarks

Black Holes

There has been voluminous literature on the subject of 'black holes' ever since the onset of the theory of general relativity in the early decades of the 20th century. It is claimed that this state of the limit of condensation of a star — its gravitational collapse to minimum size and maximum density is a consequence of the theory of general relativity. There are two models of a black hole that we discussed in Chapter 2.

The first model is the commonly-held view that follows from the vacuum solution of Einstein's equations, shown in the Schwarzschild metric (2.11). I have argued that this does not relate to a real star according to the meaning of general relativity theory. This is because the geometrical (left) side of his equation is generally interpreted as a *reflection* of the existence of matter, manifested by the energy-momentum tensor on the right side of this equation. But, with this interpretation, in the case of the vacuum, where the right side of the equation is identically equal to zero, *everywhere*, the only acceptable solution of the equation must be the flat spacetime metric of special relativity. In addition, the solution (2.11) entails a singularity in space. I have argued (and Einstein insisted upon) the fact that with the theory of general relativity, as an explanation of physical phenomena, all singularities must be excluded.

The second model of the black hole is a star not described by a singularity. It is a star that is so dense that the field of geodesics associated with it is *closed*. In this case, all signals, material and/or radiation, including the gravitational interaction, emitted by the star must propagate along its closed geodesics and then be re-absorbed by the same star. This model of the black hole then prohibits its being bound to another star, either a visible star or another black hole. This is because the binding entails the propagation of the gravitational force from the emitter to the absorber along its geodesics. Since the latter are closed, all signals emitted by this star must be re-absorbed by the same star.

The existence of this type of star is dependent on the existence of stable solutions of the general relativity equations for a family of closed geodesics. There is yet no conclusive evidence that such stable solutions exist.

Pulsars

Another type of exotic star is the 'pulsar'. This is a star that is seen to emit periodic pulses of radiation, at fixed intervals. The present-day (commonly accepted) idea is that this is a highly condensed neutron star, in rotational motion. It is a star composed of tightly bound neutrons. It emits radiation in pulses, as we would see a lighthouse emit light in pulses as it rotates.

My speculation[41] is the following: contingent on the existence of stable solutions for a family of closed geodesics constituting a 'black hole', a pulsar could be a pulsating black hole. The idea is this: any star (including a black hole) is a plasma — a sea of positively and negatively charged matter. The natural dynamics of a plasma is its periodic pulsations. For a maximally condensed star, in the black hole state it would not emit radiation to the outside world. But if it pulsates in and out of the black hole density state, periodically, then when it is out of this state, radiation would be emitted to the outside world and when it goes back into this state, radiation would not be emitted to the outside world. Thus, an outside observer would see periodic pulses from such a star, as is seen in the case of the pulsar.

On the Human Race and Cosmology

It is interesting to compare the history of the human race according to the single big-bang model of cosmology and the oscillating universe cosmology.

In the scenario of the single big bang, it appears that the human race is an accident in the cosmological scheme of the world. After the initial explosion of the big bang, when the matter of the universe started to expand and after it became sufficiently cool, individual stars and planets bound to some of them were formed. The time was then reached, as on the earth, when conditions were ripe for the formation of the human species. But this happened over a relatively short time, in comparison with the age of the universe! For example, the sun will eventually use up all of its nuclear fuel and it will then disintegrate, along with earth and its inhabitants, who would freeze out of existence and disintegrate, never to return to a human race. The remains of the human race would then join the rest of the expanding cosmic dust that constitutes the entire universe, *forever.*

In the scenario of the oscillating universe cosmology, the human race does not appear to be an accident in the cosmic scheme of things. When the expansion of any given cycle of the oscillating universe changes to a contraction, the matter of the universe, including the human race, will heat up until it fuses with all other matter into a condensed 'matter soup'. The human race will then have vaporized and become part of this maximally condensed matter that constitutes the (highly condensed) universe. The inflection point is then reached when the contraction changes to an expansion, once again, and the universe proceeds to cool down. Stars and planets will then form again, some of them with conditions conducive to the formation of a human race. Thus, in each of an indefinite number of cycles of the oscillating universe, the human race is regenerated. The human race (on earth and, most likely, intelligible species on many other planets of the universe) is then not an accident! It is just as old and ordered as the universe itself.

In the next and final chapter of this book, we will discuss some of the philosophical considerations tied to the physics of the universe that we have presented thus far.

7

Philosophical Considerations

On Truth

The primary goals of philosophy and science are the same — the truth. But there is a spectrum of the meanings of truth in philosophy and in science.

The goal of science is an understanding of nature. What the scientists do to discover the truths of nature is to first take account of the empirical facts regarding some phenomenon. With the further use of human intuition, they then formulate a set of principles that in turn acts as the *universals* that lead, by logical deduction, to particulars that are to be compared with the empirical facts. If there is a correspondence between these particulars and the empirical facts, one can then say that, *thus far*, there is an achievement of some new understanding of the phenomenon. The theory is then said to be *true* to nature. Yet this sort of truth is, in principle, refutable. The discovery of any new empirical facts that do not conform with the alleged 'true theory' or the discovery of any logical inconsistency in the formulation of the theory must then lead to a partial or total rejection of the scientific truth of the theory.

Indeed, this is the characteristic that describes our achievement of progress in the history of science — continually rejecting older ideas and replacing them with newer ideas of truth. Still it is my contention that there are threads of truth persisting from one period of 'normal science' (an existing paradigm) to the next.

Some have claimed that these periods of paradigm change are 'scientific revolutions'. This is the idea that *all* of the earlier understanding of any particular phenomenon is replaced with an entirely new understanding. I deny this claim in my assertion of the existence of the threads of scientific truth throughout the history of science. Thus, in my view, science is *evolutionary* rather than *revolutionary*.

The reason that, in principle, scientific truth must be refutable is that we can never claim the achievement of a *complete* understanding of any natural phenomenon. This was indeed Galileo's idea when he said in his Dialogues[11]: 'No matter how brilliant a theoretician may be, he (she) can never achieve a complete understanding of any natural phenomenon'. His reason for this conclusion was his belief that the *total understanding* of any natural phenomenon is unbounded in extent. Human beings are finite, thus we cannot achieve unbounded understanding; that is to say, human beings cannot be omniscient! Thus, they are bound to make mistakes along the way in searching for the truths of nature. *It is for this reason that Galileo certainly would have rejected present-day claims that physicists are on the way to achieving a theory of everything*!

In the history of science, there are special times of rapidly changing paradigms, such as the change from the classical Newtonian physics to Einstein's relativity physics in the early part of the 20th century. In this way we approach, in asymptotic fashion, scientific truths; though cognizant that we can never reach the limit of total understanding!

The concept of 'philosophical truth' is more encompassing than scientific truth. It includes the refutable sort that is scientific truth, as well as irrefutable sorts of truth, such as an analytic truth. An important example of analytic truth is a mathematical truth, such as the arithmetic statement $2+3 = 5$. Given the definition of the integers 2 and 3 as measures on a linear scale, and the definitions of the operations $+$ and $=$, the conclusion '5' is an irrefutable analytic truth (it is also called a 'necessary truth'). It is very important not to fall into the trap of confusing a scientific truth (refutable) with an analytic truth (irrefutable). The fundamental reason for this is that scientific truth relates to nature — *it is out there*. On the other hand, analytic truth is internal; it refers to a logic *invented by us*. According to the rules of this logic, a statement is true or false, *contingent on the logic that we, ourselves, have invented*! This sort of truth is not tied to nature.

It is true that the language that we use for the expression of science is mathematical. It was Galileo, among other earlier scholars, who said that 'the book of nature is written in the language of mathematics'. Thus, this language yields true or false statements, based on its own internal logic. But it is based on axioms outside of the language — the underlying concepts of the natural world. Nature is there and it is our job as scientists to probe its truths. Analytic truth, on the other hand, is based on a language, verbal or mathematical, with its inherent logic that we humans invent. That is, we cannot derive natural consequences from language — it is used in science only to facilitate an *expression* of the laws of nature.

An example in the history of science on the error of confusing analytic truth with scientific truth is the claim of the *School of Pythagoras*, in ancient Greece, that the mathematical relations between numbers must reflect the natural structure of the universe. A second example in ancient Greece was Plato's assertion that the configuration of the stars of the night sky is based on the geometrical structure of regular polygons — those whose vertices can fit onto a spherical surface. This was based on Plato's aesthetic assertion that space itself, that is to occupy matter, must be spherical since the sphere is the most perfect of all shapes! But an analytic truth, as any assertion of mathematics is based on axioms and the logic that we invent, while the physical characteristics of the universe must follow from the scientific truths of nature itself.

A third sort of truth is 'religious truth'. It is irrefutable because it is based on *faith*. An example of religious truth is the faith that a scientist has that for every physical effect in the universe there must be an underlying physical cause. The laws of nature that he (or she) seeks are indeed these cause-effect relations. Another example of religious truth is, of course, the faith that one may have in the existence of God.

We see, then, that in the scientists' investigation of the truths of the natural world, in addition to the search for scientific truth, they must resort to religious truth — their *faith* in the existence of laws of nature, before they have been established scientifically for any physical phenomenon. This is in addition to the use of analytic truth — in their use of a mathematical language (as well as a verbal language) to *express* these alleged laws of nature.

As we look up at a clear night sky, we cannot but feel awe-struck by the beauty and the order of the displays of stars, flying meteors, comets and other aspects of the universe. This quality of beauty is in addition to the scientist's attempts to understand, intellectually, what he (or she) sees with the naked eye or with the sophisticated instrumentation of modern-day astronomy. It has inspired some to proclaim that such vast order could not have come into being unless there was a God to create it. Yet this is not a scientific conclusion, subject to empirical testing, tests of logical consistency and refutability. It is a religious conclusion. Still, the scientist may not assert that it is a false claim because it is based on religious truth! It is simply that to claim a scientific truth based on a religious statement, or vice versa, is, logically, a nonsequitur because religious truth and scientific truth are in different contexts. Thus, a religious assertion is unacceptable as a *scientific conclusion. It is the reason why the US courts rejected the idea of teaching Creationism in our schools as an alternate theory of the creation of the universe — that it would be a violation of the separation between Church and State, as required in the US Constitution.*

In this regard, it is interesting to recall Einstein's comment about religion and science: 'Science without religion would be blind; religion without science would be lame.' I have paraphrased this comment by saying: 'Physics without philosophy would be blind; philosophy without physics would be lame.'

Positivism versus Realism, Subjectivity versus Objectivity

It has been my assumption that the universe is a totally ordered and closed system. Where there has been a lack of total order is in our knowledge and physical understanding of the universe. This is bound to be since we, as finite human beings, can never achieve total understanding and knowledge of the laws of the universe. This was Galileo's comment.[11]

There has been confusion among scholars between the objectivity of what *is* (the subject of ontology) and the subjectivity of what *we can know* (the subject of epistemology).

To illustrate this confusion, consider the concept of 'entropy'. This is a concept from the second law of thermodynamics. Reference is often made to the 'entropy of the universe'. If this refers to the universe as a closed system, it is a misconception. Let us first define the concept of entropy. According to the second law of thermodynamics, if a complex system is initially in a more ordered state (minimum entropy), then if left on its own it will proceed, in time, to a less ordered state. This will continue until equilibrium is reached, at maximum disorder (maximum entropy). That is, 'entropy' is a measure of the disorder of the system.

A well-known example to demonstrate this concept is from J.W. Gibbs. Insert a droplet of blue ink into a clear liquid in a beaker. Initially there is maximum order in the sense that one can specify the location of each of the molecules constituting the ink drop. As time proceeds, the ink molecules diffuse into the clear liquid until it has permanently become a pale blue color, homogeneously. That is, in this final state *our knowledge* of the location of each of the ink molecules becomes maximally disordered. This is expressed in terms of the entropy of the system becoming a maximum at the equilibrium state and remaining this way for all future times.

The question is: was this disorder (entropy) of the clear liquid plus the ink drop (objectively) intrinsic to the system? Or was this disorder a matter of our (subjective) lack of knowledge of the whereabouts of the ink molecules during the diffusion process? Did the ink molecules diffuse into the clear liquid because of the higher probability, *in our view*, of their being in the larger volume of the liquid rather than in the smaller confine of the ink drop? I don't believe that this was the *physical reason* for the diffusion process. It was rather that the forces exerted by the host liquid on the ink molecules forced them, dynamically, to move into the clear liquid.

That is to say, our *subjective knowledge* (or lack of knowledge) of the locations of the ink molecules had nothing to do with the dynamic *objective* forces that caused the diffusion of the ink molecules into the clear liquid.

The dynamical behavior of the universe as a whole — a closed system — is not influenced by any human being's knowledge or lack of knowledge of it. Thus, 'entropy of the universe', independent of our knowledge of its details, is a contradiction in terms! To claim a fundamental role of entropy in the nature of the universe is to confuse ontology (a theory of what *is*) with epistemology (a theory of what we can *know*). It is to replace objectivity with subjectivity as explanatory in our understanding of the universe. This may boost the human ego, to claim that whatever it is that we do not know does not exist, but it is certainly a false claim.

In my view, this is the same criticism that one may have for the positivistic approach of the quantum theory. To claim that the fundamental laws of matter (in the atomic domain) are laws of measurement and probability is to set subjectivity against objectivity as fundamental to the laws of nature — the laws of the universe! This claim was the view of Niels Bohr and Werner Heisenberg (the Copenhagen School) in their interpretation of the quantum theory as the most fundamental description of elementary matter. Albert Einstein and Erwin Schrödinger opposed the positivistic view of the Copenhagen School with a philosophical view of realism. This is the view that there is a real world, with inherent laws of nature for all phenomena, in any domain — from elementary particles to cosmology — the physics of the universe as a whole. That is, in the realist view the physical nature of the real world is independent of whether or not an observer has knowledge and understanding of its detailed behavior. With this view, it is the obligation of the scientist to probe this reality as far as possible in order to proceed toward increased fundamental understanding, rather than to merely describe reality.

The *description* of reality is indeed a necessary step in science,[42] but it must be followed by an *explanatory stage*, to reveal any new understanding of physical phenomena. This understanding is then the truth that is the ultimate goal of science. To know this is indeed to see that 'physics without philosophy would be blind'. Of course, there is much more to philosophy than the rules for acquiring scientific knowledge and understanding. Yet science is an intricate, interwoven part of philosophy. It is the reason for saying that 'philosophy without physics would be lame'. An important philosopher-scientist of the 19th/20th centuries who took this view was Ernst Mach.

On Mach's Influence in Physics and Cosmology

In 1970, I published an article on Mach's cosmological view.[43] This discussion is *apropos* for this monograph, especially in regard to Mach's global interpretation of the inertia of matter. Einstein has called this the 'Mach principle'. I will now discuss these ideas in the context of the physics of the universe.

I have argued that the acceptance of the continuous field concept as foundational, according to Einstein's theory of general relativity, as well as Faraday's earlier view, necessarily leads to a deterministic theory of the cosmos. With this approach, the mathematical language relates *holistically* to the 'observed — observer' as a *single, closed entity*. These arguments have shown that with the principle of relativity taken to its logical extreme, it can make no difference as to which part of an interaction is called 'observer' and which is described as 'observed'. That is to say, the 'objective' and the 'subjective' ingredients of an elementary interaction are only a matter of the frame of reference chosen for convenience to describe physical phenomena. This conclusion is, of course, in contrast with the assertion of the quantum theory.

The quantum mechanical limit

It then follows that the field solutions of this theory of the cosmos must relate to a closed entity — the 'elementary interaction' — that necessarily solves nonlinear, nonhomogeneous, partial differential equations, as it was anticipated by Einstein in his theory of general relativity. It is only after the general field language has been constructed and the solutions displayed that the asymptotic limit may be taken, corresponding to the *appearance* of coupled components as distinguishable entities.

Mathematically, the latter limit corresponds to an uncoupling of the nonlinear field formalism into linear eigenvalue equations that, in turn, give the probability calculus for the microscopic domain, where quantum mechanics applies. But in its general form, the theory is a completely objective description of matter, wherein the probability concept is only used as a device to compute average values of the properties of the material system that do indeed have a predetermined set of values.

The Mach principle

The deterministic approach to physics, in terms of a nonlinear field theory of the cosmos, as well as all smaller domains that is implied by a full exploitation of the principle of general relativity, is not compatible with Mach's epistemological stand of *positivism*. Nevertheless, the elementarity of the *interaction*, rather than the *particle*, is indeed compatible with Mach's interpretation of the inertia of matter — 'the Mach principle'.[44] With this view, the inertial mass of matter is not an intrinsic quality of any quantity of matter. Rather, it is a measure of the coupling of this matter with all other matter of the closed system, that is, in principle, the universe.

Finally, in the search for a general theory of matter and cosmology, in the context of the Mach principle, the following question must be asked: why is it that only the inertial quality of matter follows from the notion of a closed system? For if the fundamental description of a physical system is indeed 'closed', then one should expect *all* of the manifestations of interacting matter (e.g. the electromagnetic phenomena, nuclear and weak interactions, etc.) should also be incorporated within the same theory. That is to say, the logical implication of the Mach principle is the existence of a *universal interaction* and a *generalized Mach principle*. According to the latter, *all* of the alleged intrinsic properties of matter, such as magnetic moment, electrical charge, nuclear parameters, etc. are really measures of coupling between this bit of matter (in the limit where it can be viewed this way) and the entire closed system, that is in principle the universe, rather than intrinsic properties of matter. *This view exorcises all of the remnants of atomism of matter, as Mach anticipated.* It is indeed compatible with the holistic, continuous field concept. (*In this author's research program, the nearby matter, such as the components of a dense sea of particle antiparticle pairs, plays a dominant role in determining the inertial mass of elementary matter, compared with the rest of the universe — that nevertheless does contribute, though minutely.*)

The Mach principle and a unified field theory

When one follows through with the idea that the field structure of the spacetime language is a consequence of the mutual interactions throughout the universe, and identifies, with Einstein, the total interaction with the curvature of spacetime, then one is forced to the conclusion that all observable manifestations of matter are, in fact, properties of matter whose fundamental derivation correspond with the (variable) curvature of spacetime. Thus, by following the implication of the (generalized) Mach principle to its logical extreme, the conclusion is reached that if the spacetime coordinate system were not curved, i.e. flat *everywhere*, all of the observable interactions of matter would vanish identically. The actual description of the material cosmos (or any smaller domains) must then represent only an *approximation* in the local limit, for features that are sensitive to the variability (non-zero curvature) of the spacetime, *anywhere*.

The full exploitation of the Mach principle in the theory of relativity then implies that the field properties of spacetime are a *representation* of the mutual interaction of matter that comprises the closed physical system — in principle, the universe. The implication is that the field equations of general relativity are not *if-then* relations; they are rather *if-and-only-if* relations, that is to say, the metrical field equations that we have discussed in previous chapters of this monograph, whether in Einstein's original tensor form or in the factorized quaternion form, are *identities*. This means that the only acceptable solutions of these field equations are those that correspond with non-zero (matter) source fields. Thus, when the solutions are used that correspond to a matterless universe — such as the Schwarzschild problem for the derivation of the gravitational field *outside of* the sun — one must keep in mind that these solutions are only an approximation for the effects of bona fide non-zero sources of matter. This was discussed in Chapter 6 on the nature of the 'black hole' star as a consequence of the 'vacuum equation' of general relativity theory (pp. 60–63).

Clearly, the gravitational field of the sun does in fact relate to the existence of matter and its effect on other matter. If the matter of the sun should suddenly expire, there would be no gravitational effect left — the earth and her sister planets of the solar system would begin to wander off into outer space! The full use of the Mach principle in the theory of general relativity — implying that the spacetime language system is only a representation in the scientists' language for the mutual interaction of all of the matter of a closed physical system — is, in this author's view, the primary revolutionary concept that Einstein introduced with his theory of general relativity.

References and Notes

1. Some of the scholars of ancient Greece (among many others) who wrote on the subject of cosmology were Lucretius, Plato and Aristotle. The poet Lucretius wrote: *The Nature of the Universe*, Penguin Books, London, 1951, *trans.* R.E. Lantham. On Plato, see: F.M. Cornford, *Plato's Cosmology: The Timaeus of Plato*, Humanities Press, New York, 1937. On Aristotle, see: W.K.C. Guthrie, *trans.* Aristotle, *On the Heavens*, Harvard, 1939.
2. A discussion of Galileo's use of the telescope can be found in: W. Bixby, *The Universe of Galileo and Newton*, American Heritage, Harper and Row, New York, 1964.
3. Galileo, *Dialogues Concerning Two Chief World Systems, trans.* S. Drake, University of California Press, Berkeley, California, 1970.
4. M. Sachs, *Concepts of Modern Physics: The Haifa Lectures*, Imperial College Press, London, 2007, p. 9.
5. H.S. Thayer (ed.), *Newton's Philosophy of Nature*, Hafner, 1953; I. Newton, *Optiks*, Dover, New York, 1952.
6. W. Herschel, 'On the Construction of the Heavens', (1785), in M.K. Munitz (ed.), *Theories of the Universe*, The Free Press, Glencoe, Illinois, 1957, p. 264.
7. E. Hubble, *The Realm of Nebulae*, Yale, 1936, Chapter 1.
8. J. Silk, *The Big Bang*, W.H. Freeman, New York, 1980, p. 7.
9. See, for example, E. Schrödinger, *Space-Time Structure*, Cambridge, 1954, Chapter VII; M. Sachs, *Quantum Mechanics and Gravity*, Springer, Berlin, 2004, Sec. 2.7.
10. K.R. Popper, *Objective Knowledge: An Evolutionary Approach*, Oxford University Press, Oxford, 1972.
11. Galileo, Ref. 3, p. 101. In these Dialogues, its author says: ' ... there is not a single effect in nature, even the least that exists, such that the most ingenious theorists can arrive at a complete understanding of it. This vain presumption of understanding everything can have no other basis than never understanding anything. For anyone who had experienced just once the perfect understanding of one single thing, and had truly tasted how knowledge is accomplished, would recognize that of the infinity of other truths he understands nothing'.
12. M. Sachs, *Relativity in Our Time*, Taylor and Francis, London, 1993.

13. For a proof of Noether's theorem applied to field theory, see: N.N. Bogoliubov and D.V. Shirkov, *Introduction to the Theory of Quantized Fields*, Interscience, New York, 1980, 3rd edn., Sec. 1.2.
14. In A. Einstein's 'Autobiographical Notes', in P.A. Schilpp, (ed.), *Albert Einstein Philosopher-Scientist*, Open Court, La Salle, Illinois 1948, p. 81, he said: 'If one had the field equation of the total field, one would be compelled to demand that the particles themselves would *everywhere* be describable as singularity-free solutions of the completed field equations. Only then would the general theory of relativity be a *complete* theory.'
15. M. Sachs, On the Most General Form of Field Theory from Symmetry Principles, *Nature* **226**, 138 (1970).
16. M. Sachs, *General Relativity and Matter*, Reidel, Dordrecht, 1982.
17. M. Sachs, *Quantum Mechanics and Gravity*, Springer, Berlin, 2004.
18. Ref. 16, Chapter 6.
19. Ref. 16, Chapter 3.
20. A. Einstein, *The Meaning of Relativity: Relativity Theory of the Non-Symmetric Field*, Princeton, 1956, 5th edn. A.J. Adler, M.J. Bazin and M.M. Schiffer, *Introduction to General Relativity*, McGraw-Hill, New York, 1975, 2nd edn., Chapter 10.
21. Ref. 17, Sec. 2.7.
22. P.A.M. Dirac, *General Theory of Relativity*, Wiley, New York, 1975, Chapter 18.
23. R.D. Blandford, in S. Hawking and W. Israel, (eds.), *Three Hundred Years of Gravitation*, Cambridge, 1987, Chapter 8.
24. R.V. Pound and G.A. Rebka, *Physical Review Letters* **3**, 439 (1959).
25. A. Einstein, in P.A. Schilpp, (ed.), *Albert Einstein Philosopher-Scientist*, Open Court, La Salle, Illinois, 1948, p. 75.
26. Ref. 16, Chapter 6.
27. Ref. 16, Sec. 3.14.
28. Ref. 16, Sec. 6.13.
29. M. Sachs, 'On the Unification of Gravity and Electromagnetism and the Absence of Magnetic Monopoles', *Nuovo Cimento* **114B**, 123 (1999).
30. See, for example, P.A.M. Dirac, *General Theory of Relativity*, Wiley, New York, 1975, Chapters 8 and 9.
31. Ref. 17, Sec. 3.6.
32. M. Sachs, 'Considerations of an Oscillating, Spiral Universe Cosmology', *Annales de la Fondation Louis deBroglie* **14**, 361 (1989).
33. M. Sachs, *Quantum Mechanics from General Relativity*, Reidel, Dordrecht, 1986, Chapter 7.
34. Ref. 16, Sec. 6.16; Ref. 33, Sec. 7.8.
35. Ref. 16, Sec. 3.14.
36. M. Sachs, 'On the Electron-Muon Mass Doublet from General Relativity', *Nuovo Cimento* **7B**, 247 (1972).
37. M. Sachs, 'On the Lifetime of the Muon State of the Electron-Muon Mass Doublet', *Nuovo Cimento* **10B**, 339 (1972).
38. Ref. 33, p. 64.
39. Ref. 33, Sec. 7.3.

40. The following authors proposed the concept of 'delayed action at a distance': H. Tetrode, *Zeits, F. Physik* **10**, 217 (1922); G.N. Lewis, *Nat. Acad. Sci. Proc.* **12**, 22, 439 (1929); J.A. Wheeler and R.P. Feynman, 'Classical Electrodynamics in Terms of Direct Interparticle Action', *Reviews of Modern Physics* **21**, 425 (1949).

41. M. Sachs, 'A Pulsar Model from an Oscillating Black Hole', *Foundations of Physics* **12**, 689 (1982).

42. Further discussion on this topic is given in: M. Sachs, *Concepts of Modern Physics: The Haifa Lectures*, Imperial College Press, London, 2007, Chapter 1.

43. M. Sachs, 'Positivism, Realism and Existentialism in Mach's Influence on Contemporary Physics', *Philosophy and Phenomenological Research* **30**, 403 (1970). I am grateful to my late colleague, Professor Marvin Farber, for discussing these ideas with me.

44. M. Sachs, 'The Mach Principle and the Origin of Inertia from General Relativity', in M. Sachs and A.R. Roy (eds.), *Mach's Principle and the Origin of Inertia*, Apeiron, Montreal, 2003, p. 1.

Postscript

In this concluding section of *Physics of the Universe* there is presented four articles by the author: 'Physics of the 21st Century', 'Holism', 'The Universe' and 'The Mach Principle and the Origin of Inertia in General Relativity'. The fourth of these articles is from a presentation at the International Workshop on Mach's Principle (at the Indian Institute of Technology, Kharagpur, India, 6–8 February, 2002), published in the book: M. Sachs and A. R. Roy, (eds.), *Mach's Principle and the Origin of Inertia*, Apeiron, Montreal, 2003, p. 1.

These four papers discuss some of the bases of ideas developed in the text. A part of this discussion is duplicated in the text. As a pedagogue, I believe that there is a great amount of teaching value in repeating ideas, sometimes from different angles.

Physics in the 21st Century

Do we see any major paradigm changes coming in the 21st century in physics — changes of fundamental ideas that underlie the material world? My answer is: yes. It is because the leading ideas of contemporary physics are in conflict. The fundamental bases of the two revolutions of 20th century physics — the quantum theory and the theory of relativity — are both mathematically and conceptually incompatible![45] The main paradigm that has dominated 20th century physics has been that of the quantum theory. Yet the theory of relativity has given many correct predictions since its inception at the beginning of the 20th century. It must then be incorporated into all of the laws that underlie physics.

Atomism versus Holism

The initial instigation of the quantum theory was the experimental finding, in the 1920s, of Davisson and Germer, in the US, and G.P. Thomson in the UK, that the smallest particle of matter — the electron — has a wave nature, rather than the nature of a discrete 'thing'. These were the seminal experiments on the diffraction of scattered electrons from a crystal lattice. Instead of revealing the geometrical shadow of the crystal, as discrete particles would do, the scattered electrons revealed an interference pattern, as waves would yield. *This was conclusive evidence that the particles of matter are continuous waves!* Preceding this experimental finding, de Broglie had postulated the existence of 'matter waves'. Then, Schrödinger discovered the equation whose solutions were the matter waves discovered in the electron diffraction experiments.

Schrödinger initially tried to make his 'wave mechanics' compatible with the symmetry requirements of the theory of special relativity. He was not able to do this, with the requirement that the matter waves were to represent the (indestructible) electrons. Subsequently, it was Dirac who showed how to represent Schrödinger's 'wave mechanics' is a relativistic way.[46] But even this improvement was not entirely satisfactory because of the way that the particle wave was coupled to electromagnetic radiation.

To overcome this difficulty and complete the theory there had to be an extension of the Dirac theory to a different formalism, called 'quantum electrodynamics'. This generalization did lead to added predictions that were in agreement with the empirical facts, such as the 'Lamb shift' in the energy levels of hydrogen. However, this was theoretically unsatisfactory. First, there was no closed mathematical description of quantum electrodynamics. Second, the alleged solutions were constructed and displayed in terms of infinite series *that diverge*. Thus, there is the prediction here that all physical properties of elementary matter are infinite! Formal methods of renormalization to subtract off the infinities were discovered. However, as successful as this had been to match empirical facts, the scheme is not mathematically consistent.[47]

The Quantum Theory versus the Theory of Relativity

It is my judgment that the trouble with reconciling the quantum and relativity theories is indeed that these two approaches are mathematically and conceptually incompatible. A resolution of the problem might then be a paradigm change, replacing that of the quantum theory with the paradigm of the theory of relativity — *fully exploited*. This paradigm change entails our model of matter, dating back to the conflict in ancient Greece on the atomicity or the holistic, continuum basis of elementary matter.

One of the important atomists in an ancient Greece was Democritus.[48] His view was that any observable matter must be composed of many indestructible 'things' that characterize this matter. These are bodies that are free until they are brought into interaction with each other. On the other hand, those who believed in the continuum view of matter, such as Parmenides in ancient Greece,[48] saw the universe as a single, continuous, immovable entity. While his view of the universe was that of a static, continuous whole, another interpretation would say that the 'things' we experience are its multiple manifestations, not unlike the ripples of a pond. They are like the continuous, correlated modes of the entire undivided pond. The latter *holistic* view of the single continuum, being derivative of Parmenides' world, would be more akin to that of Spinoza. The former view is that of *atomism*. The latter is a view of *holism*, wherein there are no truly separate parts. *It is this change from the atomistic to the holistic view of the material universe that I see as the basic paradigm change that will be recognized in physics in the 21st century.*

The atomistic paradigm has dominated the thinking in physics since the ancient periods to the present time. The quantum theory views matter atomistically, even though one needs fields (of probability) — the 'matter fields' — to represent the laws of the elementary particles of matter (electrons, protons, quarks,...) This view is non-deterministic (the laws of the elements of matter are not predetermined, aside from the (laboratory-sized) measurements on their qualities). It rejects some of the causality in the laws of nature. The view is also subjective and linear (because probabilities necessarily obey a linear calculus).[49]

On the other hand, the theory of relativity, as a fundamental basis for our understanding of matter, is holistic, based on the continuous field concept. This view, opposing the atomistic approach, was originally introduced by Michael Faraday in the 19th century, to understand the laws of electricity and magnetism. Thereby there was introduced into physics the basic conflict between the model of matter in terms of mass points and their motions, as speculated by Newton in the 17th century, and the continuous field concept to underlie all of the laws of matter. The theory of relativity (in its special or general forms) leads to the continuous field view and holism. It will be explained in more detail in later paragraphs of this essay.[50]

When, in 1927, the empirical result was revealed that the fundamental particles of matter have a continuous field (wave) form, as shown conclusively in the electron diffraction experiments, the Copenhagen School (led by Niels Bohr) refused to give up the discrete particle view. To resolve this problem, they resorted to the concept of 'wave particle dualism'. That is to say, if an experiment is designed to see the particle as a continuous wave, it is so at that time. But if a different sort of experiment is designed to see the particle as a discrete mass point, it is so at that time. This is in accordance with the epistemological approach of *positivism*. On the other hand, the opposite view of Einstein is that of *realism* — that the qualities of the electron, as a fundamental matter component, are independent of any sort of observation that may (or may not) be made by a macro-observer.[51]

It was Max Born who saw that the Schrödinger wave mechanics may be put into the form of a probability calculus. Thus, the Copenhagen School interpreted the 'matter wave' $\psi(x, t)$ of Schrödinger's equation as related to the probability density $\psi^*\psi(x, t)$ for finding, *upon measurement by a macro-observer*, that the electron is at the location x at the time t. This paradigm, that the laws of nature are the laws of probability (i.e. laws of chance) then carried forth to the 21st century.

The alternative view of the electron diffraction experiment is that the particles of matter are indeed fundamentally continuous matter waves. (This was Schrödinger's and de Broglie's original interpretation!) It is a paradigm change that fits in well with Einstein's view that the theory of relativity — based on the continuous field concept — underlies all of the laws of nature. It is based on a totally ordered universe and holism, where probability and measurement play no fundamental role.

The Theory of General Relativity and Holism

When one fully exploits the theory of general relativity, as a fundamental theory of matter, one is led naturally to a continuum, holistic view of matter. To demonstrate this, we start with the basic axiom that underlies this theory — 'the principle of relativity'. Its assertion is this: The expressions of any law of nature, as determined by any single observer, in all possible reference frames from his or her view, must be in *one-to-one correspondence*.[52]

For example, if an experimenter studies the law of a burning candle, in his or her own frame of reference, and then compares this expression of the law in any other frame of reference relative to his or her own, such as a law of the burning candle in a moving rocket ship, the expressions of the law of the burning candle, in the relatively moving reference frames, must be the same. *This is equivalent to saying that the laws of nature must be fully objective.*

There may be objections to calling this principle a law of physics, asking: how could a law be a law, by definition, if it were not fully objective? That is, it seems that the principle of relativity is a tautology! — *a necessary truth*. It would be like asserting that 'woman is female'. Of course, this is a true statement, but it is empty — because it is only a definition of a word! But the principle of relativity is not a tautology, because it depends on two tacit assumptions that are not necessarily true in the real world. (They are only *contingently* true.) One is that there are laws of nature in the first place. That is, it is assumed by the scientist that for every effect in the real world, there is an underlying cause — *a law of total causation*. It is then the obligation of the scientist to search for the cause-effect relations — the laws of nature to *explain* the natural effects observed. The search for such explanation is the raison d'etre of the scientist. But this law of total causation is not a necessary truth of the world. It is a law that is based on the scientist's faith in its truth.

The second tacit assumption that underlies the principle of relativity is that we can comprehend and express the laws of nature. This is where the space and time parameters come into the picture. They form a language (not the only possible language) that is invented for the purpose of facilitating an expression of the laws of nature.

The Continuous Field Concept

If we assume that the spacetime language for the laws of nature is a continuous set, then the principle of relativity requires that these laws must be preserved in form (covariant) under the *continuous transformations* from one reference frame to any other continuously connected frame. *(In principle, the spacetime language could be discrete rather than continuous. Then the laws of nature would be in terms of difference equations (governed by the rules of arithmetic) rather than differential equations (governed by the rules of calculus). Nevertheless, in accordance with all of our discoveries in science thus far, the spacetime language is continuous.)* Thus, the solutions of the laws of nature must be continuous functions of the spacetime coordinates. These functions are the continuous fields that underlie the true nature of matter. For example, in the case of electromagnetism, these are the solutions of the Maxwell field equations (or their factorized spinor version, to be discussed later on). A further feature, based on the requirement of the inclusion, in the local limit, of conservation laws (of energy, momentum and angular momentum) is that these fields must be analytic — continuously differentiable to all orders. *(This is based on Noether's theorem.[53])* That is, it follows from the principle of relativity and the assumption that the space and time parameters is a continuous set, that the laws of nature, including the laws of conservation of energy, momentum and angular momentum, are field equations whose solutions are *regular* — continuous and analytic, *everywhere. (This is a requirement that Einstein emphasized throughout his study of the theory of (special and general) relativity.)*

The Language of General Relativity

According to relativity theory, the spacetime language is used to facilitate an expression of the laws of matter. Since matter fields are generally continuously variable, it follows that the metric of spacetime must entail continuously variable coefficients. That is to say, the differential invariant metric is:

$$ds^2 = \Sigma_{\mu\nu} g^{\mu\nu}(x) dx_\mu dx_\nu = ds'^2$$

where the sum over μ and ν is taken over the four space and time coordinates. This is a Riemannian geometrical system.

The principle of relativity then requires that the laws of nature must be covariant (form preserving) under the same set of continuous transformations that leave ds invariant, i.e. $ds \rightarrow ds' = ds$.

The metric tensor $g^{\mu\nu}(x)$ is a continuous, regular function of the four spacetime coordinates x. Because it is a symmetric tensor, $g^{\mu\nu} = g^{\nu\mu}$, it has ten independent components that reflect the material content of the closed system, that in principle is the universe. The *Einstein field equations* are then a set of ten nonlinear differential equations that determine the metric tensor components, given the material content of the system, as represented by the energy-momentum tensor of the matter, as its source.[54] It was this formalism that *explained* the phenomenon of gravity, superseding Newton's theory of universal gravitation, which *described* rather than explained gravitation in the experimentation before the 20th century. Einstein's field theory predicted all of the phenomena covered by Newton's (atomistic) theory, in addition to extra effects that were not predicted by Newton's theory.

The asymptotic limit of the Riemannian geometrical system, as the matter content is depleted toward zero (a vacuum, *everywhere*) is the Euclidean geometrical system, with $g^{00} \rightarrow 1$, $g^{kk} \rightarrow -1$, ($k = 1, 2, 3$) and $g^{\mu \neq \nu} \rightarrow 0$. Thus, in special relativity the metric is: $ds^2 = (dx^0)^2 - (dx^1)^2 - (dx^2)^2 - (dx^3)^2$. The geodesics of the Riemannian spacetime are variable curves. The family of these curves is a *curved spacetime*. The family of geodesics of the Euclidean geometry is a set of straight lines. This is a *flat spacetime*. Thus, the flat Euclidean spacetime is the ideal limit of a matterless system — a vacuum *everywhere*. Thus, the theory of special relativity, based on the Euclidean spacetime, is only true, in principle, for the idealistic limit of a vacuous universe, *everywhere*. However, special relativity may be used as a good approximation for the theory of general relativity where the actual geodesics are curves, but approximated by straight line paths in particular regions. The significance of the geodesic is that it is the natural path of the unobstructed body. That is, the path of a body on a curve due to the action of an external potential in a Euclidean space is equivalent to its natural motion in a curved Riemannian space. *This statement is the essence of the principle of equivalence.*[55]

A Unified Field Theory

According to the theory of general relativity, one does not have different domains where one type of force or another, such as electromagnetism or gravity, is in effect, while the other is not. There are no sharp boundaries in the field theory of this approach. The implication is that the continuous field solutions of these physical laws incorporate all of the possible forces that matter exerts on matter, as well as the inertial properties of this matter. It is rather that one type of force or another will dominate the others under particular physical conditions. But all forces that matter exerts on matter are present at all times — the long-range electromagnetic and gravitational forces as well as the short range forces called 'weak' and 'nuclear'. The next question is: what does the general form of such a unified field take that is logically required of the theory of general relativity?

In one of his papers on the unified field theory, in 1945,[56] Einstein advised that one should not only pay attention to the geometrical logic of the laws of physics, but also pay attention to its algebraic logic. The latter refers to the underlying algebraic symmetry group, and its representations. What I have found is that the underlying group of the theory of relativity is a Lie group — a set of continuous, analytic transformations that project the field equations of physics, in any reference frame, to any *continuously connected* reference frame. That is, this is a *continuous group*, without any reflections in space or time. I have called this 'the Einstein group'. What I have found is that the *irreducible representations* of this group obey the algebra of quaternions.[57] These representations, in turn, have as their basis functions the (two-component) spinor variables. The asymptotic limit of the representations of the Einstein group of general relativity is the set of representations of the 'Poincaré group' of special relativity.

Thus, the algebraic symmetry group of relativity theory tells us that the most primitive fields that solve the laws of nature are the spinor and quaternion variables, mapped in a curved spacetime for general relativity or the flat spacetime for special relativity. Indeed, the reason for Dirac's relativistic generalization of Schrödinger's wave mechanics to a spinor formalism in special relativity is that the imposed symmetry was that of relativity theory, not that spin is a uniquely quantum mechanical property. That is to say, *any theory in physics that is to conform with the symmetry requirements of (special or general) relativity theory* — from elementary particle physics to cosmology — must, in its most primitive (irreducible) form, be in terms of spinor field solutions.

With the quaternion-spinor formalism in the curved spacetime, it is found that the ten relations of Einstein's field equations, that have already provided an explanation for gravity, and superseded Newton's theory of universal gravitation, *factorize* to sixteen field relations whose solutions are the quaternion variables $q^\mu(x)$. These are the four quaternion components of a four-vector. Thus, this variable has $4 \times 4 = 16$ independent components. The new factorized field equations then replace Einstein's ten tensor field equations, as the fundamental representation of the spacetime language of general relativity. The factorization essentially occurs because of the elimination of the reflection symmetry in space and time in Einstein's field equations, which was not required in the first place, since the covariance is defined in terms of a continuous group alone (the 'Einstein group').

It was then shown that the sixteen field equations could be separated, by iteration, into ten equations that are in one-to-one correspondence with the form of Einstein's tensor equations, to explain gravity, and six equations that are in one-to-one correspondence with the form of Maxwell's equations, to explain electromagnetism. Thus, this quaternion factorization of Einstein's field equations yields a formalism that unifies gravity and electromagnetism, in terms of the single sixteen-component quaternion field $q^\mu(x)$ — *this is the unified field theory sought by Einstein.*[58]

A further feature of this formalism is that the geodesic equation takes a quaternion form, predicting that this is a set of four independent equations, rather than one. That is, the 'time' that parameterizes the path of a body in the curved space is defined here as a set of four parameters rather than one, as in the usual geodesic equation. (*This result is in agreement with the speculation of William Hamilton, in the 19th century, that the quaternion number system, which he discovered, would turn out to most fundamentally represent the time measure in the problems of physics.*[59])

A further implication of the quaternion expression of the field laws was that Maxwell's field equations for electromagnetism also factorizes to a pair of two-component spinor field equations. It follows here, as it did in the factorization of Einstein's tensor formalism, from the elimination in the fundamental field equations of the reflection symmetry in time and space for electromagnetic interactions. A generalization that then occurs is that, in addition to the standard scalar electromagnetic interaction, there is predicted to be a pseudoscalar electromagnetic interaction. (*A prediction of the latter was that of the Lamb shift in the hydrogen atom, calculated to be in better empirical agreement with the data than the standard quantum electrodynamics.*) There is also indicated here a basis for the 'weak interaction' in the nuclear domain, as following from this generalization to spinor-quaternion form of the electromagnetic field equations.[60]

The Elementary Particle Domain

Summing up, the geometrical and algebraic logic of the theory of relativity predicts that the laws of nature must be field equations in terms of spinor and quaternion variables in a curved spacetime that unify the laws of gravity and electromagnetism. The Einstein symmetry group, when taken to its logical extreme, predicts there are no fundamental 'spin one' particles. Thus, the 'photon' of the electromagnetic theory is not an elementary particle. Rather, it is a mode of the continuum that carries the electromagnetic interaction at the speed of light c, from one (spinor) component of charged matter to another. This is the long-range electromagnetic (scalar) interaction, as shown in the binding of the electron and proton to form the hydrogen atom. The predicted (short-range) pseudoscalar electromagnetic interaction (that must accompany the scalar interaction because of the lack of reflection symmetry in the underlying field) entails the spinor form of the electromagnetic field, as a neutrino field that carries the binding of the electron and proton to form the neutron. Thus, the neutron is not an elementary particle. It is a composite of electron, proton and the spin one-half electromagnetic field of coupling between them that we associate with the neutrino field.

Similarly, it has been found in this research program applied to the elementary particle domain that the pions, that mediate the short-range (strong) nuclear interaction (as seen, for example, between protons and neutrons), are *composites* of fundamental particle fields. Indeed, the only elementary particle fields here are the four stable matter fields: electron, positron, proton and antiproton. The photon and the neutrino are *virtual fields* that affect the coupling between the stable fields that make up the composite elementary particle matter fields — as modes of a continuum. The investigation shows that the numerical values of the ratio of the masses of charged to neutral pions and the ratio of their lifetimes is in conformity with the empirical data. The composite model of the kaon also yields the correct ratio of CP violation to non-violation compared with the empirical data.

Quantum Mechanics from General Relativity

One of Einstein's anticipations for the future of physics was that the formal expression of quantum mechanics would follow from a closed form field theory of matter, rooted in the theory of general relativity. That is, he believed that the asserted foundations of the quantum theory — a probability calculus, indeterminism, partial causality and the role of measurement — are false. He believed that the quantum theory appears to have these characteristics because it is an incomplete expression of the laws of matter. This is analogous to the incompleteness of statistical mechanics to describe fully the dynamics of a many body system. Then how is it that quantum mechanics is a very accurate expression of the laws of elementary matter in the atomic and elementary particle domain, especially at non-relativistic energies?

What I have found is that Einstein was right about this. In the fully exploited theory of general relativity, the spinor-quaternion formalism leads to the full (Hilbert space) expression of quantum mechanics as a linear approximation for a nonlinear field theory of the inertia of matter, that is rooted in general relativity.[61] That is, in this view, the 'quantum phenomena' come from a general field theory as an approximation — that is to say, it is an *incomplete* description of the atomic domain. The *completed* expression is indeed a non-atomistic, holistic field theory of the inertia of matter, based on the continuous closed system in general relativity. It is generally deterministic and nonlinear. Statistics and measurement by macro-observers do not play any fundamental role, as they do in the Copenhagen view of quantum mechanics. The full derivation of this result is given in my books.[59,60,61]

From the Inertial Masses of Elementary Particles to Cosmology

On mass

The quaternion-spinor factorization in general relativity leads to the formal expression of the quantum mechanical equations. This, in turn, leads to an explicit relation between the masses of elementary matter and the features of the curved spacetime.

Without going into the mathematical details here, the derivation revealing this connection of inertial mass to spacetime is as follows: one starts with the most primitive expression where the inertial mass appears in the physics of elementary matter. This is the quantum mechanical equations in special relativity, with the Majorana form of two coupled two-component spinor equations in the Euclidean spacetime. On the left side of one of these spinor equations is a quaternion operator (defined in terms of the Pauli matrices and the unit matrix as the basis elements of the quaternion) acting on one type of spinor. On the right hand side of this equation is the second type of spinor that is a reflection of the first, multiplied by the mass parameter m. The second spinor equation is a reflection of the first.[62] *(The combination of these two (two-component) spinor equations yields the (four-component) Dirac bispinor equation that in turn restores reflection symmetry. The latter, in turn, in the nonrelativistic limit, gives back the Schrödinger wave equation).*

To derive the mass, one first sets the right side of this equation equal to zero. One then re-expresses the left side of the equation in a curved spacetime. The constant Pauli matrices and the unit matrix, σ^μ, (the basis elements of the quaternion) then become the quaternion *field* $q^\mu(x)$, and the ordinary derivatives are replaced by the covariant derivatives. The latter introduces an extra term called 'spin-affine connection', Ω_μ in the generalization: $\sigma^\mu \partial_\mu \rightarrow q^\mu(\partial_\mu + \Omega_\mu)$. The spin-affine connection is a necessary term in a curved spacetime, in order to make the spinor solutions *integrable*. It is the term $q^\mu \Omega_\mu$ that then leads to an explicit form for the inertial mass field (with imposed gauge invariance on the spinor formalism).

In this way the quantum mechanical formalism was derived from the Riemannian spacetime with the quaternion-spinor expression and, as a byproduct, the inertial mass field is derived from first principles in a generally relativistic theory of the inertia of elementary matter.

An important consequence of the relation between the inertial mass of elementary matter (say, an electron) and the spin-affine connection of the curved spacetime is that as the rest of the matter of a closed system (in principle, the universe) tends to zero, i.e. to a vacuum, *everywhere*, the spin-affine connection, and therefore the mass of the given particle, tends to zero. This is a prediction that is in accord with the *Mach principle*.

An important further prediction is that the masses of elementary (spin one-half) particles of matter occurs in doublets. These are the eigenvalues of the two-dimensional mass field that depends on

$q^\mu \Omega_\mu$. A calculation yielded the ratio of masses of the electron to its heavy sister to be $2\alpha/3$, in a first approximation, where $\alpha \approx 1/137$ is the fine structure constant. Thus, the ratio $m(\text{electron})/m^*(\text{electron})$ is close to $1/206$. This is the empirical ratio of the mass of the electron to that of the muon. The physical reasons for the appearance of the higher mass value is that the background of any given particle of matter, say an electron, is a dense sea of particle-antiparticle pairs, in their (derived) ground state of null energy and momentum. When the given electron couples to one of the pairs of this background, the pair excites and the spin-affine connection thereby changes, which in turn alters the mass of this electron. It is determined further that the excited pair in the vicinity of the given electron decays to its ground state in the order of 10^{-6} seconds, thereby restoring the heavy electron to its original minimal value. This lifetime is indeed the order of the empirically measured lifetime of the muon. Thus, it has been found that the muon is explained in general relativity as a higher mass state of the electron mass doublet. The physical characteristics of the muon are identical to those of the electron, except for its inertial mass and stability.

The prediction then also follows that the proton has a heavy sister, whose mass is the order of 193 Gev. It is interesting that recent experimentation has identified the 'top quark' with a mass that is close to this. A major difference, however, is that the quark is a fractionally charged particle while the heavy proton is integrally charged.[63] Experimentation on the fractional, or integral charge of this particle would then be a good test of the validity of the 'standard model' of the quantum theory versus the source of inertial mass from general relativity.

Summing up, the quaternion-spinor formalism in general relativity leads to the full (Hilbert space) formal expression of quantum mechanics as a linear approximation for a generally covariant, non-linear field theory of the inertia of matter. An important prediction here is that the elementary particle called 'muon' is explained in general relativity as the heavy member of an electron mass doublet. A second important prediction is that the mass of an elementary particle vanishes as the other matter of a closed system that is in its environment tends to zero. This is in agreement with the statement of the *Mach principle.*

On cosmology

There are several new implications of the quaternion formulation of general relativity in the problems of cosmology. 1) Galaxies rotate as a natural motion, as it is seen in astronomical observations. 2) The plane of polarization of cosmic radiation must rotate as it propagates through the cosmos. This is the *Faraday Effect. This effect has been seen in astronomical observations.* 3) The dynamics of the universe as a whole is oscillating between expansion and contraction. Its configuration is spiral rather than isotropic.[64] 4) The universe is populated with a dense sea of particle-antiparticle pairs, each in a (derived) ground state of null energy, momentum and angular momentum. This medium could serve as the dark matter that is called for in the researches of astrophysics. 5) This prediction concerns the seemingly uneven distribution of matter and antimatter in the universe. The prediction in this theory gives the following scenario: at the beginning of the expansion phase of each cycle of the oscillating universe (at the 'big bangs') a fraction of the bound particle-antiparticles ionize into positively charged particles and negatively charged particles, as they rotate with the spiraling matter of the universe. These oppositely charged particles of matter thereby give rise to enormous magnetic fields of opposite polarity. Thus, the effect of the magnetic fields competes with the spiral rotation of all of the matter of the universe. The ionized matter is then separated into matter (positively or negatively charged) and antimatter (negatively or positively charged), sent in opposite directions. It then follows that while our region of the universe is populated predominantly with matter, there are regions of the universe populated predominantly with antimatter and their complex configurations: complex systems of nuclei, atoms, molecules and larger forms made up of bound antiparticles. Thus, matter is separated from the antimatter in different regions of the universe.[65]

Holism and Realism

As it has been emphasized earlier, it is my belief that the major paradigm change in 21st century physics will be from *atomism to holism*. The epistemology will change from *positivism to realism*. Both of these changes come from the replacement of the quantum theory with the theory of relativity as the basis of the laws of physics.

The holistic view implies there are no sharp boundaries between individual 'things'. That is, what we think of, from our subjective perceptions, as independent things, are really correlated modes of a single, continuous entity that in principle is the universe. It is opposite to the atomistic view of matter that has dominated Western thinking for thousands of years.

The view of *positivism*, proposed by the Copenhagen inter-pretation of the quantum theory, has led to a 20th century belief in the fundamental importance of the subjective element in our *definition* of matter and the world. In my view, it is an egocentric (pre-Copernican) innovation that will fade in the 21st century. I believe that it will be replaced by a restoration of the view of the world in terms of a totally objective entity, whether or not we humans are there to perceive all (or even a tiny fraction) of its ramifications!

References

45. These conflicts are discussed in detail in: M. Sachs, *Einstein versus Bohr*, Open Court, La salle, Illinois, 1988, Chapter 10.
46. See: P.A.M. Dirac, 'The Evolution of the Physicist's Picture of Nature', *Scientific American* **208**, 45 (1963).
47. The difficulty is discussed in: P.A.M. Dirac, 'The Early Years of Relativity', in: G. Holton and Y. Elkana (*eds.*), *Albert Einstein: Historical and Cultural Perspectives*, Princeton University Press, 1982, p. 79.
48. T.V. Smith (*ed.*), *From Thales to Plato*, University of Chicago, Phoenix Books, Chicago, Illinois 1956. Democritus, p. 39; Parmenides, p. 15.
49. The philosophy of the quantum theory is discussed in: M. Sachs, *Concepts of Modern Physics: The Haifa Lectures*, Imperial College Press, London, 2007, Chapters V, VI.
50. See A. Einstein, *Ideas and Opinions*, Crown publishers, New York, 1994, 'The Fundamentals of Theoretical Physics', p. 337; 'Relativity and the Problem of Space', p. 398.

51. Wave-particle dualism is discussed in ref. 1 and in: M. Sachs, 'On Wave Particle Dualism', *Annales de la Fondation Louis de Broglie* **1**, 129 (1976).
52. M. Sachs, *Relativity in Our Time*, Taylor and Francis, London, 1993, Chapter 2.
53. See, for example, N.N. Bogoliubov and D.V. Shirkov, *Introduction to the Theory of Quantized Fields*, Interscience, New York, 1980, 3rd edn., Sec. 1.2.
54. The relation between geometry and matter is expressed in the Einstein field equations. It is derived in: M. Sachs, *General Relativity and Matter*, Reidel Publishing Co., Dordrecht, 1982, Chapter 6.
55. M. Sachs, 'On the Logical Status of Equivalence Principles in General Relativity Theory', *British Journal for the Philosophy of Science* **27**, 225 (1976).
56. A. Einstein, 'Generalization of the Relativistic Theory of Gravitation', *Annals of Mathematics* **46**, 578 (1945). A sequel to this paper is by A. Einstein and E. Straus, *Annals of Mathematics* **47**, 731 (1947).
57. Ref. 54, Chapters 3, 6.
58. The unified field theory is demonstrated explicitly in Ref. 54 and in: M. Sachs, *Quantum Mechanics and Gravity*, Springer, Berlin, 2004, Chapter 3.
59. Ref. 54, p. 63.
60. M. Sachs, *Quantum Mechanics and Gravity*, Springer, Berlin, 2004, Sec. 5.5.
61. M. Sachs, *Quantum Mechanics from General Relativity*, Reidel Publishing Co., Dordrecht, 1982, Chapter 9.
62. Ref. 54, Chapter 4.
63. M. Sachs, 'A Proton Mass Doublet from General Relativity', *Nuovo Cimento* **59A**, 113 (1980); M. Sachs, 'Interpretation of the Top-Quark Mass in Terms of a Proton Mass Doublet in General Relativity', *Nuovo Cimento* **108A**, 1445 (1995).
64. Ref. 60, Sec. 8.7.3.
65. Ref. 60, Sec. 8.9.

Holism

A Leading Paradigm Change in 21st Century Physics

What will be the most significant paradigm change in 21st Century Physics? In my opinion it is the holistic model of matter, in its replacing the atomistic model.

The debate between the ontologies of holism versus atomism as underlying the fundamental nature of matter goes back to ancient Greece, in the Western Civilization and to the ancient teachings in Buddhism and Taoism in Asia. In ancient Greece, the view of atomism was that of Democritus and his disciples and the holistic continuum view was that of Parmenides and his disciples. The atomistic model of Democritus has predominated to the present time until it was challenged, first by Michael Faraday and his field concept in the 19th century and then by Albert Einstein in his formulation of the theory of general relativity, based on the continuous field concept. The full implementation of the field concept, in turn, implies the holistic ontology as a fundamental basis for matter.

In the 20th century, the continuous field concept came into different branches of physics. Faraday's field concept was successfully applied to the laws of electricity and magnetism, as formulated by James Clerk Maxwell in the form of Maxwell's equations for electromagnetism. It was this continuum field that was to be the potential force exerted by charged matter on other charged matter.

The field concept then appeared in the expression of the quantum theory, in the form of a continuously distributed field of probability.

Thirdly, the geometrical field came into physics with the appearance of the theory of general relativity. Here, there is a continuously distributed field that is a measure of the metrical properties of spacetime. It is this field that predicted the gravitational manifestation of matter. This theory then superseded Newton's theory of universal gravitation. It replaced the discrete particularity of matter (atomism) with the continuous field of force; it replaced the concept of 'action at a distance' with the concept of propagating forces at a finite speed, between interacting fields of matter.

Newton's atomistic theory of matter was modified by a different particle theory of micromatter — the quantum theory. It is represented in terms of the matter fields that are probability distributions — described by the wave function ψ. In this view, the matter components of a system are still atomistic, and uncorrelated, unless an interaction between them might be 'turned on'.

The field theory of general relativity that superseded Newton is not atomistic. It is based on the interpretation of the matter fields as (Spinozist) modes of a single continuum. This is the entire universe. All of the (infinite number of) distinguishable modes are correlated fields, just as the multiple ripples of a pond are, in fact, correlated modes of the single pond. Further, each of the multitude of matter fields, $\psi_1(x), \psi_2(x), \ldots$ is mapped in the same four-dimensional spacetime, x. This is a view of holism.

In the quantum mechanical particle view, each of the matter fields of the system is mapped in its own spacetime, $\psi_1(x_1), \psi_2(x_2), \ldots$. Thus, for an N-particle system, one must deal with a $4N$-dimensional spacetime. Schrödinger referred to this N-particle system in quantum mechanics as 'entanglement'. It is because each of the N matter fields intertwines. In the holistic ontology there are N fields but only one spacetime, thus there is no entanglement!

Summarizing, the continuum, non-singular field model of matter in general relativity is holistic and non-atomistic. What appear as separated things — electrons and protons, people, planets, stars, galaxies, etc. — are really correlated modes of a single continuous entity — the universe. It is a model of the material universe that exorcises all remnants of atomism. All of these modes are dynamically coupled. This reflects the spirit of the Mach principle, originally applied only to the inertial mass of matter.

Quantum Mechanics is a contemporary theory in physics that also claims a holistic ontology, but in a different context to the holism of a continuous field theory. This is because of the Copenhagen School that insists that the measurement process by macro-observers of micro-matter is an integral part of the *definition* of elementary matter. That is, the 'observer' and the 'observed' are holistically a single entity.

But this is not really holism. The 'observer' in Quantum Mechanics obeys a different set of rules than the 'observed'. The former obeys the rules of classical physics while the latter obeys the rules of quantum physics. In this view, the only reality is the set of responses of a classical 'observer' to the quantum mechanical 'observed'. Indeed, this is a true approach of *positivism*. The philosophy of holism, on the other hand (in the sense of Spinoza), is an approach of *realism*.

Finally, let me comment on the extension of the holistic approach from physics to human relations. The following is taken from my article: 'The Influence of the Physics and Philosophy of Einstein's Relativity on my Attitude in Science: An Autobiography', in M. Ram (ed.), *Fragments of Science: Festschrift for Mendel Sachs* (World Scientific, Singapore, 1999), p. 201.

The holistic view that is logically implied in our understanding of matter by Einstein's theory of general relativity has interesting consequences when it is applied to the subject of human relations. The implication is that in reality there are no individual, separable things — protons, people, trees, planets, stars, galaxies, and so on. Instead these are the multitude of distinguishable manifestations of the whole, single, continuous entity that is the universe.

With this view, social relations must be viewed holistically. We are not separate egos, separate nations and separate ethnic groups. We are all correlated in terms of the whole entity, a continuum of which we are only some of its infinitesimal manifestations. To understand this concept is to understand that in the long run we cannot gain for ourselves by being destructive to others or to our environments or to any aspect of all of nature. This is because we are all modes of a single holistic entity. By analogy, if a man has a sore toe, it would not be helpful to cut it off! To solve the problem must then require a cure for the toe as an integral part of the whole body. [This is the idea of holistic medicine, which has been practiced by many different groups of the human society for many centuries — in the Orient, by the Native Americans, by the aboriginal peoples throughout the world, and so on.]

With this view, the concept of war cannot be a resolution for any political problems between nations. Nor can ethnic or social prejudice be taken seriously as a reasonable conclusion — for the best interests of any individual or any societal group.

I believe that if the community of physicists eventually sees the truth in this holistic view, which has in large part been demonstrated in the physics of our time in the theory of general relativity and other aspects of modern physics, it could infuse into our culture and affect our attitudes toward all of the social and environmental relations that we encounter. In my view, such a philosophical attitude would certainly be for the betterment of the human race.

The Universe

Understandably the human race has been fascinated over the ages with the universe. From the periods of ancient Greece and Asia, a primary pursuit has been the observations of the stars and planets of the night sky — the subject of astronomy — and the speculations to understand them.

Greek Astronomy[66]

In ancient Greece, about 2,300 years ago, Aristotle argued that the Earth is at rest at the center of the universe, and all of the heavenly bodies are in rotation about it. In that period, Aristotle's idea was confirmed by the observations of the astronomer Ptolemy. This is the 'geocentric model' of the universe. A few centuries before Aristotle, Pythagoras speculated on the structure of the heavens, based on mathematical relations that he discovered. Some of these relations were found, in part, from his discovery of irrational numbers and the relations of the frequencies of vibration of a stretched string to the fractions of the length of the string under vibration. In this, Pythagoras saw a connection between physics and mathematics. That is, he reasoned that the discovery of mathematical relations must imply physical relations in the real world. (*Pythagoras did not personally record his own findings. They were recorded by his disciples — the 'School of Pythagoras'.*)

Aristotle's teacher, Plato, also speculated about the distribution of stars, based on his idea of 'forms' and symmetry of the heavens.

He argued that the stellar objects must be at the vertices of regular solids — these are geometrical forms that can be inscribed with their vertices on the surface of a sphere — which he assumed was the shape of space. (*These are the five regular solids: cube, tetrahedron, icosahedron, octahedron and dodecahedron.*) Plato believed that the space of the universe must be spherical because the sphere is the most perfect of forms.

On the subject of the extent of space, Aristotle believed that it must be finite. He argued that the only logical reason for space to exist is that it must be there to occupy matter. From his observations, he deduced that the amount of matter in the universe is finite, thus he concluded that space must be finite.

Aristotle agreed with Plato that there are four primary elements — air, earth, fire and water. He then asserted that earth must be 'down'. He concluded that material objects would fall to Earth because their natural place is 'down'. (*One might see this assertion as Aristotle's law of gravity.*)

Aristotle deduced that there is an absolute center of the universe. This is the location of our planet Earth. All other stellar objects must then be in circular motion about Earth, at the center of the universe (in agreement with Ptolemy's astronomical observations).

It should be mentioned that Plato's understanding of the heavens was qualitatively different from that of his pupil, Aristotle. Plato's view was abstract, based on the ideas that we form in our minds. The reality of the world must then be deduced by rational analyses from these impressions on our minds. On the other hand, Aristotle's view of the world was concrete — that the ways of the world are how we directly experience it. (*This difference in understanding the world is indicated in a painting, in the Renaissance period, of Raphael, entitled, 'School of Athens'. In this painting, Plato is seen pointing upward to the heavens; Aristotle is seen pointing downwards, to Earth.*)

As the years progressed to the medieval times, the scholar of the Roman Catholic Church, Thomas Aquinas, argued that Aristotle's ideas of the universe were compatible with the Holy Scriptures. Thus, the Christian Church adapted most of Aristotle's views as God's truth. An exception of disagreement was Aristotle's conclusion that there was no beginning of time. In contrast, the Church advocated, in accordance with the Biblical Scriptures, that there was a beginning of time, when God created the universe, *ab initio*. In the 15th century, contrary to the views of the Church, Copernicus discovered from his observations of the night sky that the Earth moves relative to the Sun — that Earth is not at the center of the universe! (*Thus, that the human being, residing on Earth, is not at the center of the universe!*) He theorized that the Sun is indeed at rest at the center of the universe, with all heavenly bodies (including Earth) revolving about it. This is the 'heliocentric model' of the universe.

Galileo's Physics[67]

In the 16th century that followed Copernicus, Galileo was the first astronomer to use the telescope, to magnify his observations of the heavens. He agreed with Copernicus that the Earth moves, but he argued further that motion, *per se*, is a subjective feature of our observations. Thus, from his understanding, it is just as true to say that the Sun moves relative to the Earth, *from the Earth's perspective*, as it is to say that the Earth moves relative to the Sun, *from the Sun's perspective*. It was his assertion that it is the physical law that binds the Earth and the Sun that must remain unchanged in form, independent of the perspective taken. This is called 'Galileo's principle of relativity'. It is a very important precursor for 'Einstein's principle of relativity', that logically underlies his theory of general relativity, that was to come in 20th century physics. At the present stage of the history of physics, the latter is an important underlying law of the physics of the universe as a whole — the subject of cosmology.

Modern-day Astronomy

In the 16th century, Galileo believed that the display of stars of the night sky, that we call the 'Milky Way', is the entire universe. It was learned centuries after Galieo that the 'Milky Way' is only one of its galaxies. A galaxy contains a very large number of stars; it is one of an infinitude of other galaxies of the universe. Our Sun is an average-sized star among a very large number of constituent stars of the 'Milky Way'. It was discovered in the 19th century that 'Milky Way' has a neighboring galaxy, called 'Andromeda', that forms a binary system with 'Milky Way'.

With the present-day high-resolution instrumentation (such as the Hubble telescope), we have now gained a great deal more information about the night sky. There are exotic stellar objects, such as the pulsars and quasars. Pulsars are extremely dense, small stars that emit periodic bursts of radiation. Quasars are the most distant stellar objects we see, with enormous emitted radiant energy.

Recent high-resolution telescopic observations reveal the shapes and dynamics of the galaxies. Most are 'pancake' shaped, bulging with their constituent stars at their centers; most have spiral arms. Our own Sun is in one of the spiral arms of 'Milky Way'. Some of the galaxies have the shape of ellipsoids. (*It has been my speculation that the galaxies behave like plasmas and that they therefore change their shapes during their natural pulsations. In this view, there is a possibility that there are continual transformations between the spiral and ellipsoidal shaped galaxies.*)

It has been seen that the 'flat' spiral galaxies rotate about an axis that is perpendicular to their two-dimensional forms. Originally, it was thought that this rotation is due to the gravitational pull on them by their neighboring galaxies, but the masses of the galaxies are known as well as their mutual separations. Using the approximation of Newtonian gravity, it was then seen that the neighboring galaxies would be inadequate to cause the observed rotations of the galaxies. It was then speculated that there must be some invisible matter (to us) that permeates the universe that is responsible for the rotations of the galaxies. This unseen matter is called 'dark matter'. Candidates for this would be a dense sea of (non-zero mass) neutrinos and antineutrinos, or a dense sea of particle antiparticle pairs. Such electrically neutral matter couples gravitationally to other matter.

The Expansion of the Universe and the Hubble Law

In the 1920s Edwin Hubble discovered that the universe is expanding — the galaxies of the universe are moving away from each other at an accelerating rate. The empirical finding was that the speed of any galaxy relative to another, v, is linearly proportional to their mutual separation R, i.e., $v = HR$. The constant of proportionality, H, is called Hubble's constant. Its determination then allows us to extrapolate backward in time to see when the expansion started — the 'big bang'. It turns out to be the order of 15 billion years ago.

Hubble's law was established from the Doppler shift of radiation emitted by one galaxy that is moving away from another. In the visible spectrum, the frequencies of monochromatic radiation of the emitting galaxy are then shifted toward the red end of the spectrum. Thus, it was Hubble's conclusion that all of the galaxies of the universe are moving away from all other galaxies — the universe is expanding.

This expansion does not mean that the universe as a whole is moving into empty space. There is no empty space outside of the universe — the universe is all there is! What the expansion signifies is that, from any observer's view, the density of matter at any point in the universe is decreasing with respect to his or her time measure.

Extrapolating backward in time, as we have indicated above, leads to the time when the density of matter was at a maximum and it was maximally unstable. This state led to a gigantic explosion — the 'big bang' — starting the presently observed expansion.

The Spiral Universe

It is usually assumed that the 'big bang' resulted in an expanding *isotropic* and *homogeneous* distribution of the matter of the universe. *There is no underlying reason to believe these assumptions.*

Indeed, astronomical observations of the night sky with high-resolution instrumentation, (such as the Hubble telescope) or even with the naked eye, reveals that the distribution of matter in the universe, the galaxies and other stellar configurations is not isotropic or homogeneous. Galaxies cluster in some domains of the sky and not in other domains. Nor is the distribution of galaxies and other stellar matter the same in all directions of observation.

The question then arises: if the theory of general relativity is to provide a dynamical theory of the universe, as a closed system, are there solutions of its field equations that yield a non-isotropic and non-homogeneous matter distribution? In my research program, I have found that under reasonable approximations, there are solutions of the (generalized, quaternion form) of the general relativity field equations that represent a spiraling universe.[68] These solutions are the 'Fresnel integrals'. This dynamic is characterized by two inflection points, one where matter is maximally dense and there ensues an expansion of the matter of the universe, (the 'big bang'), and the second where the expanding matter has reached sufficient rarefaction that the expansion changes to a contraction. In the former phase, the gravitational forces are predominantly repulsive; in the latter phase, the predominant gravitational forces are attractive. Thus, from these solutions it is predicted that the universe continually expands and contracts in a spiral configuration, in cycles. If the presently estimated time since the last big bang, which is the order of 15 billion years, is approximately a half-cycle of the oscillating universe, it may be estimated that the period of a cycle is the order of 30 billion years. That is, every 30 billion years, from our time frame, there is another 'big bang'.

In this view, then, the universe is oscillating between expansion and contraction in spiral fashion. The matter of the universe is then neither isotropic nor homogeneous, as it is believed to be in present-day astrophysics. *The spiral configuration has many other appearances in other domains of Nature!*

In the foregoing analysis, the Hubble law is derived as a first approximation for a covariant law of the expansion and contraction of the universe. It is clear that the original expression of the Hubble law, $v = HR$, is not covariant. That is, any continuous spacetime transformation from the reference frame of an observer, where this law is seen to hold, to any other reference frame, would change its form. Nevertheless, this noncovariant form, that is empirically correct, is found to be an approximation for a truly covariant law of the dynamics of the matter of the universe. The spiral universe leads to this covariant law, in conformity with the theory of general relativity.

Matter and Antimatter in the Universe

An interesting question in contemporary elementary particle physics that may perhaps be resolved in the context of the spiral universe cosmology, is this: why is it that most elementary particles in our region of the universe are matter, such as electrons and protons? The bulk matter that we experience, including our own bodies, is made up of composites of these particles, and their binding in terms of photon and neutrino fields.

We have detected in experimentation, from nuclear accelerators and in cosmic rays, antiparticles in our region of the universe, such as positrons (positively charged electrons) and antiprotons (negatively charged protons). But why is this antimatter not freely abundant in our region of the universe? A possible scenario that answers this question comes out of my research program that entails the spiral, oscillating universe cosmology.

Pair Annihilation and Creation

From my research program, there is an exact solution of the nonlinear field equations for the particle antiparticle bound pair (electron-positron or proton-antiproton) that represents its true ground state.[69] The energy, linear momentum and angular momentum are all null for this state of the pair. Thus, when the particle and antiparticle are bound in this state, that is $2\,mc^2$ units of energy below the state where they would be free of each other; if this quantity of energy would be supplied to such a bound pair it would dissociate, giving rise to the appearance of a free particle and antiparticle. The former bound state, at null energy, would correspond to 'pair annihilation' — but without actually annihilating the pair. It is still there, and capable of interacting gravitationally with other matter. (In this bound state, the pair would be invisible to an observer.) The latter state of dissociation would correspond to 'pair creation' — though without actually creating a pair from a vacuum! Also, in the ground state of the pair, its dynamics reveals two oppositely polarized currents, in a plane perpendicular to the direction of interaction with a detecting apparatus. This is precisely what is seen, and interpreted as the creation of two oppositely, circularly polarized photons, when a pair is said to 'annihilate'.[70]

It was then concluded that there is no reason for there not to be a very dense gas of such pairs, in their ground states, in any region of the universe. It was found that the inertial mass of an elementary particle is indeed determined by such a dense gas of particle-antiparticle pairs — yielding, for a determined density value, the correct values of inertial masses of the electron and the muon.[71] Further, a sea of such bound pairs could serve as the 'dark matter' that permeates the universe, as evoked by the astrophysicists.

The Separation of Matter and Antimatter in the Universe

The scenario for the separation of matter from antimatter in the universe, at the inflection point where the expansion ensues (the 'big bang') is as follows: after the onset of the expansion phase of the oscillating universe, in the spiral motion of any given cycle, and after some cooling has taken place, the gravitational field of the universe delivers about 1 Mev units of energy to each of a number of electron pairs, that is, $2mc^2$ to dissociate them (out of a much larger number of such pairs) and 2 Gev ($2Mc^2$) to each of a number of proton pairs, to dissociate them (where m is the electron mass and M is the proton mass).

The released particles and antiparticles are then in rotational motion of the spiraling universe. The rotating particles and antiparticles, in a plane perpendicular to the axis of rotation of the universe, being oppositely charged, create magnetic fields, parallel and antiparallel to the axis of rotation of the spiraling universe. Thus, there is a competition between the gravitational field of the matter of the universe, inducing particles and antiparticles to move in a single rotational motion of the spiraling universe, and the magnetic fields separating the directions of motion of the particles and antiparticles. More particles than antiparticles will then move in one direction and more antiparticles than particles will move in the opposite direction. Matter and antimatter then become separated at the early stages of the expansion phases of each of the cycles of the oscillating universe.

Thus, certain regions of the universe become populated with mostly matter and other distant regions of the universe become populated with mostly antimatter. Future experimental studies of the distant regions of the universe may reveal the predominance of antimatter over matter in the formation of antimatter atoms and molecules, and the more complex structures made up of composites of antimatter. One may even speculate that (in the mode of science fiction!) there are human beings in those regions of the universe composed mainly of antimatter, living on planets and breathing the air composed of antimatter.

References

66. For a general discussion of Greek astronomy, see: A.C. Crombie (ed.), *Scientific Change*, Heinmann, London, 1963; M.K. Munitz (ed.), *Theories of the Universe*, The Free Press, Glencoe, Illinois, 1957.
67. G. Galilei, *Dialogue Concerning the Two Chief World Systems*, University of California Press, 1976, transl. S. Drake.
68. M. Sachs, *Quantum Mechanics and Gravity*, Springer, Berlin, 2004.
69. M. Sachs, *Quantum Mechanics from General Relativity*, Reidel Publishing Co., Dardrecht, 1986, Chapter 7.
70. C.S. Wu and I. Shaknov, *Physical Review* **77**, 136 (1950).
71. M. Sachs, *Nuovo Cimento* **7B**, 247 (1972); *Nuovo Cimento* **10B**, 339 (1972).

The Mach Principle and the Origin of Inertia from General Relativity*

There has been a great deal of discussion during the 20th century on the possible entailment of the Mach principle in general relativity theory. Is it a necessary ingredient? Additionally, there has been the question of the origin of the inertia of matter in general relativity — does inertia originate from the foundation of general relativity theory as an underlying theory of matter? I wish to demonstrate that indeed both of these features of matter are intimately related to the conceptual and mathematical structures of the theory of general relativity.

The Theory of General Relativity

The first thing that we must do, then, is to clearly define terms. What do we mean by 'the theory of general relativity'? I should like to preface this discussion with the comment that the title of the theory of relativity should be: 'the theory of general relativity' (or the 'theory of special relativity') rather than the more commonly used title: 'the general theory of relativity', (or the 'special theory of relativity') since it is the 'relativity' that is general (or special) and

*The author thanks Roy Keys for permission to republish this paper from M. Sachs and A.R. Roy (eds.), *Mach's Principle and the Origin of Inertia*, Apeiron, Montreal, 2003.

not the theory! There is indeed one theory of relativity, whether it is in the 'special' or the 'general' form, based on the single 'principle of covariance' (also known as the 'principle of relativity'). The adjectives 'special' or 'general' refer to the types of relative motion of the frames of reference in which the laws are to be compared from the perspective of any one of them. When the relative motion is inertial, we have special relativity and when it is generally nonuniform, we have general relativity. Thus, it is the 'relativity' that is special or general, not the theory — which is a single concept!

The 'principle of covariance' is the underlying axiom that defines this theory. It is an assertion of the objectivity of the laws of nature, asserting that their expressions are independent of transformations to any frame of reference in which they are represented, with respect to any arbitrary observer's perspective (frame of reference). This implies an entailment of all possible frames of reference; thus, it implies that any real system of matter is a *closed system*. Of course, when the coupling between any local component of the closed system is sufficiently weakly coupled to the rest of the system, say to the rest of the universe or to any smaller subsystem of matter, then one may use the mathematical approximation in which all that there is to represent, mathematically, is the localized material system. In the next approximation, the rest of the system could perturb it. But for the actual *unapproximated* closed system, the implication is that there is no singular, separable 'thing' of matter. Any constituent matter is always relative to other components that together with it makes up the entire closed system — not as singular 'parts', but rather as the modes of a single continuum.

In fundamental terms, then, the principle of covariance implies, ontologically, a *holistic model*, wherein there are no individual, singular, separable things; the closed system is rather a single system without independent parts! This is also an implication of the definition of the inertial mass of matter, according to the *Mach principle*, which we will discuss in detail later on.

The model of matter we have come to, then, from the principle of covariance of the theory of general relativity, is one of *holism*. What we observe as individual separable 'things', that we call 'elementary particles' or 'atoms' or 'people' or 'galaxies', are really each correlated modes of a single continuum. The peaks of these modes are seen to move about and to interact with each other. But indeed they are not independent, separable things, as they are all correlated through the single matter continuum, of which they are its manifestations. This single continuum is, in principle, the universe.

The Mach Principle

We have seen that the qualities of localized matter, such as the inertial mass or electric charge of 'elementary particles', are really only measures of their interactions within a closed system of matter, between these entities and the rest of the system. Thus, their values are dependent, numerically, on the rest of the matter of the closed system, of which they are elementary, inseparable constituents. Their masses and electric charges are then measures of coupling within a closed system, not intrinsic properties of 'things' of matter. The dependence of the inertial mass of localized matter, in particular, on the rest of the matter of the 'universe', is a statement of the Mach principle.

It should be emphasized, however, that what Mach said about this was not the commonly stated definition of the principle. The latter is the assertion that only the distant stars of the universe determine the mass of any local matter. In contrast to this, in his *Science of Mechanics*,[72] Mach said that *all* of the matter of the universe, *not only the distant stars*, determines the inertial mass of any localized matter.

I have found in my research program in general relativity, that the primary contribution to the inertial mass of any local elementary matter, such as an 'electron', are the nearby particle-antiparticle pairs that constitute what we call the 'physical vacuum'. (The main developments of this research are demonstrated in my two monographs: *General Relativity and Matter*,[73] and *Quantum Mechanics from General Relativity*[74].) A prediction of this research program is that the main influence of these pairs on the mass of, say, an electron comes from a domain of the 'physical vacuum' in its vicinity, whose volume has a radius that is the order of 10^{-15} cm. Of course, the distant stars, billions of light-years away, also contribute to the electron's mass, though negligibly, just as the Sun's mass contribution to the weight of a person on Earth is negligible compared with the Earth's influence on this person's weight! Nevertheless, it was Mach's contention that in principle *all* of the matter of the closed system — the nearby, as well as far away constituents — determines the inertial mass of any local matter.

Newton's Third Law of Motion

I believe that the first indication in physics of the holistic view of a closed material system came with *Newton's third law of motion*. I see this law as a very important precursor to the holistic aspect of Einstein's theory of relativity. The assertion of this law is that for every action (force) exerted on a body A, by a body B (that is located somewhere else), there is an equal quantity of (reactive) force exerted by A, oppositely directed, on B. According to this law of motion, then, the minimal material system must be the two-body system A-B. A *or* B, as individual, independent 'things', then loses meaning since, with this view, the limit in which A (or B) is by itself an entity in the universe does not exist!

One other mathematical feature (not noted by Newton) that is implied by his third law of motion is that the laws of motion of matter must be fundamentally *nonlinear*. For if A's motion is caused by a force exerted on it by B, which in turn depends on B's location relative to A, then the reactive force exerted by A on B, according to Newton's third law of motion, causes B to change its location relative to A. Consequently B's force on A changes. Thus, A's motion would be changed from what it was without the reactive force on B. We must conclude, then, that A's motion affects itself by virtue of the intermediate role that is played by B in the closed system, A-B. The mathematical implication of this effect is in terms of *nonlinear laws of motion* for A (as well as for B). Thus, we see that, at the foundational level, the model of matter, even in Newton's classical physics of 'things', must be in terms of a closed system that obeys nonlinear mathematical laws of motion.

The Generalized Mach Principle

As we will see later on, the principle of covariance of the theory of general relativity implies that the basic variables of the laws of matter must be continuous, nonsingular fields, *everywhere*. The laws of matter must then be a set of coupled, nonlinear field equations for all of the manifestations of the closed continuum that in principle is the 'universe'. Thus, we see here how the Mach principle is entirely intertwined with the theory of general relativity, regarding the logical dependence of the inertial mass of local matter on a closed system.

The theory of general relativity goes beyond the Mach principle. It implies that *all* of the qualities of local matter, not only its inertial mass, are measures of dynamical coupling between this 'local' matter and the rest of the closed material system, of which it is a constituent. I have called this 'the generalized Mach principle'.[73] Thus, the foundational aspects of the theory of general relativity imply an ontological view of holism wherein all remnants of 'atomicity' are exorcised. With this view, the 'particle' of matter, as a discrete entity, is a fiction. What these 'things' are, in reality, are manifestations (modes) of a single matter continuum.

Let us now discuss the role of space and time in general relativity theory. We will then go on to show how the inertial mass of elementary matter emerges from the field theory of general relativity. Finally, it will be seen how the formal expression of quantum mechanics (the Hilbert space formalism) emerges as a linear approximation for a nonlinear, generally covariant field theory of the inertia of matter.

The Role of Space and Time in Relativity Theory

The assertion of the *principle of covariance* entails two scientific (i.e. in principle, refutable) assertions. One is the existence of laws of all of nature. This is the claim that for every effect in nature there is a logically connected cause. This assertion is sometimes referred to as the 'principle of total causation'. These relations between causes and effects are the laws of nature that the scientists seek.

The second implication of the *principle of covariance* is that the laws of nature can be comprehended and expressed by us. This is, of course, not a necessary truth. But *as scientists*, we have faith in its veracity. The *expressions* of the laws of nature are where space and time come in. In this view, space and time are not entities in themselves. Rather, they provide the 'words' and the logic of a language, invented for the sole purpose of facilitating an expression of the laws of nature. *It is important to know that the concepts entailed in the laws of nature underlie their language expressions — in one expression or another.*

The space and time parameters and their logic then form an underlying grid in which one *maps* the field solutions of the mathematical expressions of the laws of nature. The logic of the languages of the laws of nature is in terms of geometric and algebraic relations, as well as topological relations in some applications. With the assumption that a space and time grid forms a continuous set of parameters, the solutions of the laws of nature are then continuous functions of these parameters. These are the 'field variables'. They are the solutions of the 'field equations', field relations continuously *mapped* in space and time. According to the *principle of covariance*, the field equations must maintain their forms when transformed to *continuously connected* spacetime frames of reference.

It might be mentioned here, parenthetically, that there is no logical reason to exclude a starting assumption that the language of spacetime parameters is a discrete, rather than a continuous grid of points. In this case, the laws of nature would be in the form of difference equations rather than differential equations. However, the implications of the spacetime parameters as forming a continuum, in the expressions of the laws of nature, as continuous field equations, agrees with all of the empirical facts about matter that we are presently aware of. Thus, we assume at the outset that the spacetime language is indeed in the form of a continuous set of parameters. Its geometrical logic in special relativity is Euclidean and in general relativity it is Riemannian. The algebraic logic is in terms of the defining symmetry group of the theory of relativity; it is a *Lie group* — a set of continuous, analytic transformations. The reason for the requirement of analyticity of the transformation group will be discussed in the next paragraph. The Lie group in special relativity is the 10-parameter *Poincaré group*; in general relativity, it is the 16-parameter *Einstein group*.

A requirement of the spacetime language, stressed by Einstein, as mentioned above, is that the field solutions of the laws of nature — the solutions of the 'field equations' — should be *regular*. This is to say, they should not only be continuous, but also analytic (continuously differentiable to all orders, without any singularities) *everywhere*. I am not aware that Einstein gave any explicit reason for this requirement in his writings. However, I believe that it can be based on the *empirical requirement* that the (local) flat spacetime limit of the general field theory in a curved spacetime must include laws of conservation — of energy, linear momentum and angular momentum. For, according to *Noether's theorem*,[75] the analyticity of the field solutions is a necessary and a sufficient condition for the existence of these conservation laws. Strictly, there are no conservation laws in general relativity because, covariantly, a 'time rate of change' of some function of the spacetime coordinates in a curved spacetime cannot be separated from the rest of the formulation that can go to zero. Thus, the *laws of conservation* apply strictly to the local domain. The conservation laws are then a local limit of global laws in general relativity. In the latter global field laws, a time rate of change can no longer be separated, by itself, from a four-dimensional differential change of functions mapped in a curved spacetime. That is to say, in the curved spacetime the continuous transformations of a purely time rate of change of a function of the space and time coordinates, from its frame of reference where it may appear by itself, to any other continuously connected frame of reference, leads to a mixture of space and time differential changes. In this case we cannot refer to an *objective* conservation (in time alone) of any quantity, in the curved spacetime.

Thus, we see that, based on the foundations of the theory of general relativity, we have a closed, nonsingular, holistic system of matter. It is characterized by the continuous field concept wherein the laws of nature are expressed in terms of nonlinear field equations that maintain their forms under transformations between any continuously connected reference frames of spacetime (or other suitably chosen) coordinates. Their field solutions — the 'dependent variables' — are *regular functions* of the space and time parameters, that is to say, they are continuous and analytic (nonsingular) *everywhere*. The space and time parameters and their logical relations form the language of the 'independent variables' in which the field variables are mapped. The generalized Mach principle is then a built-in (derived) feature of this holistic field theory in general relativity.

Inertia and Quantum Mechanics from General Relativity

Thus far I have argued that the (generalized) Mach principle is automatically incorporated in the (necessarily) holistic expression of the theory of general relativity, as a general theory of matter. I now wish to show how, in particular, the inertial mass of matter enters this theory of matter in a fundamental way. I will try to avoid, as much as possible, the mathematical details of this derivation. They are spelled out in full in my two monographs.[73,74]

In my view, the revolutionary and seminal experimental discovery about matter that relates to the basic nature of its inertia was made 82 years ago, when it was seen that, under particular conditions, particles of matter, such as electrons, have a wave nature. These were the experimental discoveries of electron diffraction by Davisson and Germer, in the US, and independently, by G.P. Thomson, in the UK.[76] What they observed was that electrons scatter from a crystal lattice with a diffraction pattern, just as the earlier observed X-radiation does. The 'interference fringes' of the diffraction pattern emerge when the momentum, p, of the electron is related to the *de Broglie wavelength* $\lambda = h/p$, where h is Planck's constant, and the magnitude of p is such that λ is the order of magnitude of the lattice spacing of the diffracting crystal. (This relation between a (discrete) particle variable — its momentum p — and a (continuous) wave variable — its wavelength λ — was postulated by Louis de Broglie, three years before the experimental discovery.[77])

The (discrete) particle, electron, was discovered 25 years earlier by J.J. Thomson (the father of G.P. Thomson) is his cathode ray experiments. Yet, the conclusion about the discreteness of the electron from the cathode ray experiment was indirect. This is because one never sees a truly discrete object (in any observation)! What one sees, such as in J.J. Thomson's experiment, is a localized, but *slightly smeared* 'spot' on the phosphorescent face of the cathode ray tube. One then extrapolates from this 'spot' to the existence of an actual discrete point where the electron is said to land on the screen. Nevertheless, a close examination of this *smeared spot* would reveal that inside of it, there is indeed a diffraction pattern! Thus, another possible interpretation of the experiments whereby one thinks that one is seeing the *effects of* a discrete particle is that what is actually seen is a 'bunched' continuous wave — that there is no discrete particle in the first place!

The discovery of the wave nature of the electron was a momentous and revolutionary discovery for physics. It signified a possible *paradigm change* in our ontological view of matter, from the *atomistic*, particularistic model held since the ancient times, to a continuum, *holistic* model. In the former view, macroscopic matter is viewed as a collection of singular, elementary bits of matter that may or may not interact with each other to affect the physical whole. In contrast, in the continuum, holistic view, there is a single continuous matter field. What is thought of as its individual constituents is in this view a set of manifestations (modes) of this continuum, that is, in principle, the universe! These manifestations may be electrons, or trees or human beings or galaxies. They are all correlated aspects of a single continuum — they are *of* its infinite set of modes, rather than things *in it*.

In the 1920s, when the continuous wave nature of the electron was discovered, the physics community was unwilling to accept this paradigm change, from particularity to holism and continuity of the material universe. Instead, mainly under the leadership of N. Bohr, M. Born and W. Heisenberg (the *Copenhagen school*), they opted to declare a philosophical view of *positivism*. The view was to assert that if an experiment, using macroscopic equipment should be designed to look at micromatter, such as the electron, as a (discrete) particle, as in the cathode ray experiment, this is what the electron would be *then*. But if a different sort of experiment were designed to look at the electron as a (continuous) wave, as in the electron diffraction study, this is what it would be *under those circumstances*. In other words, the type of measurement that is made on it by a macroscopic observer determines the nature of the electron (or any other material elementary particle), even though the continuous wave and discrete particle views logically exclude each other! This *positivistic* epistemological concept claims that all that can be claimed to be meaningful is what can be experimentally verified at the time a measurement is carried out. Thus, it is said that both the 'wave' aspect and the 'particle' aspect of the electron are true, though in different types of measurements. This is called 'wave-particle dualism'. It is the basis of the theory known as 'quantum mechanics', that was to follow for describing the domain of elementary particles of matter.

Inertial Mass from General Relativity

The *correspondence principle* has been an important heuristic in physics throughout history. I now wish to use this principle to show that the most primitive expression of the laws of inertial mass can be seen in a generalization in general relativity of the quantum mechanical equations in special relativity. We will then extend the quantum mechanical equations in special relativity to derive the field equations for inertia in general relativity.

   The equations we start from are the *irreducible* form of quantum mechanics in special relativity — the two-component spinor form (called the Majorana equations). This is irreducible in terms of the underlying symmetry group of special relativity — the *Poincaré group*. The latter is a set of only continuous transformations (i.e. without any discrete reflections in space or time) that leave the laws of nature covariant in all inertial frames of reference, *from the perspective of any one of them.* It is the following set of two coupled two-component spinor equations (units are chosen with $h/2\pi = c = 1$):

$$(\sigma^\mu \partial_\mu + I)\eta = -m\chi \tag{1a}$$

$$(\sigma^{\mu *}\partial_\mu + I^*)\chi = -m\eta \tag{1b}$$

To restore reflection covariance, one may combine the two spinor field equations (1ab) to yield the single four-component Dirac equation in terms of the bispinor solution, where the top two components are $(\eta + \chi)$ and the bottom two components are $(\eta - \chi)$.

But the more primitive form of the quantum mechanical equations in special relativity, based on the irreducible representations of the underlying *Poincaré symmetry group* — a continuous group without reflections — is in terms of the coupled two-component spinor equations (1ab).

In the wave equation (1a), I is the interaction functional that represents the dynamical coupling of all other matter components of the closed system to the given matter field (η, χ), in accordance with the (generalized) Mach principle. $\sigma^\mu \partial_\mu$ is a first order differential operator, $\sigma^\mu = (\sigma^0; \sigma^k)$, where σ^0 is the unit 2-matrix and σ^k ($k = 1, 2, 3$) are the three Pauli matrices. (The set of four matrices σ^μ correspond with the basis elements of a quaternion.) Thus, the operator $\sigma^\mu \partial_\mu$ is geometrically a scalar, but algebraically, it is a quaternion. I^* is the time reversal (or space inversion) of I and $\sigma^{\mu *} = (-\sigma^0; \sigma^k)$ is the time reversal of σ^μ.

The spinor field equations (1ab) are the *irreducible* form of the quantum mechanical equations in special relativity. In the limit as $v/c \to 0$, where v is the speed of a matter component relative to an observer and c is the speed of light, these equations (and the four-component Dirac equation) reduce to the nonrelativistic Schrödinger equation for wave mechanics.

Our goal is to *derive* the inertial mass of matter m from a theory of matter in general relativity. This is instead of inserting m into the equations, later to have its numerical values adjusted to the data, as it is done in the conventional formulation of quantum mechanics in special relativity. We accomplish this by 1) setting the right-hand sides of equations (1ab) equal to zero and 2) globally extending the left-hand sides of these equations to their covariant expression in a curved spacetime.

Regarding the latter step, we extend the ordinary derivatives of the spinor fields to *covariant derivatives* as follows:

$$\partial_\mu \eta \rightarrow (\partial_\mu + \Omega_\mu)\eta \equiv \eta_{;\mu} \tag{2}$$

where Ω_μ is the 'spin-affine connection' field. It must be added to the ordinary derivative of a two-component spinor in order to make the spinor field (η, χ) integrable in the curved spacetime. Its explicit form is:

$$\Omega_\mu = (1/4)(\partial_\mu q^\rho + \Gamma^\rho_{\tau\mu} q^\tau) q^*_\rho$$

where $\Gamma^\rho_{\tau\mu}$ is the ordinary affine connection of a curved spacetime.[73] The quaternion field $q^\mu(x)$ is defined fundamentally in terms of the invariant quaternion metric of the spacetime, $ds = q^\mu dx_\mu$ of the (factorized) Riemannian (squared) differential metric invariant, $ds^2 = g^{\mu\nu} dx_\mu dx_\nu$. The quaternion field q^μ is a 16-component variable that is, geometrically, a four-vector, but each of its components is quaternion-valued. It was found to be a solution of a factorized version of Einstein's field equations. It replaces the metric tensor $g^{\mu\nu}$ of Einstein's formalism.[73] The quaternion q^*_ρ is the quaternion conjugate (time-reversal) to q_ρ.

Thus, with $m = 0$ and the global extension of the left-hand side of Eq. (1a) as indicated above, the matter field equation becomes:

$$q^\mu(\partial_\mu + \Omega_\mu)\eta + I\eta = 0$$

Transposing terms we then have:

$$(q^\mu \partial_\mu + I)\eta = -q^\mu \Omega_\mu \eta \tag{3}$$

If the explicit inertial mass is to be derived from first principles in general relativity, then using the correspondence principle, compared in the special relativity limit with $m\chi$ in Eq. (1a), it must come from the spin-affine connection term on the right side of Eq. (3).

Indeed, a mathematical analysis showed that there is a *mapping* between the time-reversed spinor variables as follows[73]:

$$q^\mu \Omega_\mu \eta = \lambda[\exp(i\gamma)]\chi \qquad (4)$$

where $\lambda = (1/2)[|\det\Lambda_+| + |\det\Lambda_-|]^{1/2}$ is the modulus of a complex function and $\gamma = \tan^{-1}[|\det\Lambda_-|/|\det\Lambda_+|]^{1/2}$ is its argument, where $\Lambda_\pm = q^\mu \Omega_\mu \pm$ h.c., 'h.c.' stands for the 'hermitian conjugate' of the term that precedes it and 'det' is the determinant of the function.

Finally, applying the requirement of gauge invariance to the field theory, with the gauge transformations:

first kind: $\eta \to \eta \exp(-i\gamma/2)$, $\chi \to \chi \exp(i\gamma/2)$

second kind: $I \to I + (i/2)q^\mu \partial_\mu \gamma$,

the phase factor in Eq. (3), (using Eq. (4) on the right-hand side) is automatically transformed away. The field equation (3) — the global extension in general relativity of Eq. (1a) — then takes the form:

$$(q^\mu \partial_\mu + I)\eta = -\lambda\chi \qquad (5a)$$

Its time-reversed equation (the global extension of (1b)) is:

$$(q^{\mu*} \partial_\mu + I^*)\chi = -\lambda\eta \qquad (5b)$$

Gauge covariance is a necessary and sufficient condition for the incorporation of the laws of conservation in the field laws, in the asymptotically flat spacetime limit of the theory. Thus, the empirical facts about the existence of conservation laws of energy, linear and angular momentum in the (asymptotically flat) special relativity limit of the theory, dictate that gauge covariance is a necessary symmetry, in addition to the, continuous group symmetry in general relativity (the 'Einstein group') of the field theory.

We see, then, in using the correspondence principle, comparing the generally covariant field equations (5ab) with the asymptotically flat special relativity limit (1ab), that the function λ plays the role of the inertial mass of matter, m. Thus, we may interpret the generally covariant equations (5ab) as the defining field relations for the inertial mass of matter.

As we asymptotically approach the flat spacetime limit, equations (5ab) approach equations (1ab) and the generally covariant solutions (η, χ) approach the flat spacetime elements of the Hilbert function space $\{\eta_1, \ldots \eta_k, \ldots ; \chi_1, \ldots \chi_k, \ldots\}$, with the condition of square integrability (and normalization) imposed on these spinor variables. In this (Hilbert space) limit of the formalism, the expectation values of the positive-definite field λ is the set of squared eigenvalues (the mass spectrum formula):

$$\lambda_k^2 = |\langle \eta_k| (-q^{\mu*} \Omega_\mu^+)_a (q^\mu \Omega_\mu)_a |\eta_k\rangle|$$

where the subscript 'a' denotes the asymptotic value of the term in parentheses as the flat spacetime limit *is approached*, and the 'dagger' superscript denotes the 'hermitian conjugate' function.

A few points about the inertial mass field λ should be noted. First, in the actual flat spacetime limit, the spin-affine connection Ω_μ vanishes so that in this limit $\lambda_k = 0$. The vanishing of the spin-affine connection field occurs only for the vacuum — the absence of all matter, *everywhere*. Thus, the derivation from general relativity of the vanishing of the inertial mass $\lambda_k = 0$, where there is no other mass to couple to, is in accordance with the statement of the Mach principle.

A second important point is that, as the modulus of a complex function, λ is positive-definite. This implies that any macroscopic quantity of matter, being made up of these 'elementary' units of matter with positive mass, must itself have only positive mass. The implication is that, *in the Newtonian limit of the theory*, the gravitational force has only one polarization. It is either under all conditions repulsive or under all conditions attractive. In view of the locally observed attractive Newtonian gravitational force, it must then under all circumstances be attractive. This conclusion is in agreement with all of the empirical data on Newton's force of gravity. It has never been derived before from first principles, either in Newton's classical theory of gravitation or in the tensor formulation of Einstein's theory of general relativity. This result implies that in the Newtonian limit of the theory there is no antigravity, i.e. no gravitational repulsion of one body from another.

The Oscillating Universe Cosmology

In the generally curved spacetime of the theory of general relativity, the role of the gravitational force is not directly related to the mass of matter, as it is in Newton's theory. As we see in the geodesic equation in general relativity (the equation of motion of a test body) the 'force' acting on a body relates to the 'affine connection' of the curved spacetime. The latter is a *non-positive-definite* field. Thus, the general prediction here is that under particular physical circumstances (of sufficiently dense matter and high relative speeds between interacting matter), the 'gravitational force' can be repulsive. Under other physical circumstances (of sufficiently rarefied matter density and low relative speeds of interacting matter), the gravitational force can be attractive.

This result in general relativity, applied to the problem of the universe as a whole, implies an oscillating universe cosmology. At one inflection point, the matter components of the universe begin to repel each other, dominating the attractive components of the general gravitational force, thence leading to the *expansion phase* of the universe, with the matter continuously decreasing its density. Then, when the matter of the universe becomes sufficiently rarefied, and relative speeds between interacting matter are sufficiently low, another inflection point is reached where the attractive component of the gravitational force begins to dominate and initiates the *contraction phase* of the universe. This continues with ever-increasing matter density until the conditions are again ripe for the repulsion of matter to dominate. The universe then reaches the inflection point once again for a turnaround from contraction to expansion. The expansion phase starts again, until the next inflection point, when the attractive force takes over once more, and so on, *ad infinitum*.

The answer to the question: how did the matter of the universe get into the maximum instability stage at the last 'big bang' (the beginning of the present cycle of the oscillating universe) is then: before the last expansion started, the matter of the universe was contracting toward this physical stage. This view of the oscillating universe denies the idea of a mathematical singularity at the inflection point — at the *beginning* of any particular cycle of the oscillating universe — that is commonly believed by present-day cosmologists who adhere to the 'single big bang' model.

This cosmology also rejects the present-day model wherein there is an absolute time measure — the 'cosmological time' — measuring the time since the last big bang happened. The latter view of absolute time is incompatible with Einstein's theory of relativity, wherein there is no absolute time measure. It is replaced in relativity theory with a totally covariant description of the universe wherein the time measure (as the space measure) is a function of the reference frame from which it is determined. The universe itself cannot be expressed in terms of an absolute reference frame. In the theory of relativity, there are no absolute frames of reference or time measures.

Summary

I have argued that the basis of the theory of general relativity implies that any material system is necessarily a *closed system*. This, in turn, implies a *holistic model* of matter, whereby there are no separable, individual particles of matter. It is a view that is compatible with a continuum, rather than in terms of a collection of discrete particles of matter.

The first empirical evidence for this continuum view was the discovered wave nature of matter in the experiments on electron diffraction in the 1920s. The 'matter waves' (as they were named by their discoverer in theory, Louis de Broglie) may then be viewed as the infinite number of *correlated* manifestations (modes) of a single continuous whole — in principle, the universe. This implies that the inertial mass of any local matter is not intrinsic, but rather it is dependent on all of the other matter of the closed universe (the Mach principle). It also follows that *all other* physical properties of matter, as well as inertial mass, such as electric charge, are not intrinsic, but also measures of coupling within the closed system of matter. This is the 'generalized Mach principle'.

It was seen that the formal expression of quantum mechanics in special relativity relates, by means of a *correspondence principle*, to a generally covariant field theory of inertia, in general relativity. The formal expressions of quantum mechanics in special relativity, in accordance with the irreducible representations of the *Poincaré group*, are a set of two coupled, two-component spinor equations. Each is a time reversal (or space reflection) of the other. The mass parameter is conventionally inserted in a way that appears as a mapping between the two sorts of (reflected) spinors. Removing the mass term in the latter expression, and globally extending the rest of the equation to a curved spacetime, based on the symmetry of the *Einstein group* of general relativity, leads to the covariant field theory of inertial mass. With the added symmetry of gauge invariance, the field equation is recovered with a *mass field* appearing where the mass parameter was initially inserted.

The asymptotic limit toward the flat spacetime of the latter (nonlinear) field equations in general relativity for inertia is the formal structure of (linear) quantum mechanics — as a linear approximation. This analysis has then led to the derivation of quantum mechanics (the Hilbert space structure) as a linear approximation for a generally covariant field theory of the inertia of matter.

References

72. E. Mach, *The Science of Mechanics*, Open Court, La Salle, 1960, p. 267, *transl.* T.J. McCormack.
73. M. Sachs, *General Relativity and Matter*, Reidel Publishing Co., Dordrecht, 1982.
74. M. Sachs, *Quantum Mechanics from General Relativity*, Reidel Publishing Co., Dordrecht, 1986.
75. Noether's theorem is derived in E. Noether, *Goett. Nachr.* **235** (1918). It is explicated further in: C. Lanczos, *The Variational Principles of Mechanics*, Toronto, 1966, Appendix II and in R. Courant and D. Hilbert, *Mathematical Methods of Physics*, Interscience, New York, 1953. The application to fields is demonstrated in N.N. Bogoliubov and D.V. Shirkov, *Introduction to the Theory of Quantized Fields*, 3rd edition, Interscience, New York, 1980, Sec. 1.2.
76. C.J. Davisson and L.H. Germer, *Phys. Rev.* **30**, 705 (1927), G.P. Thomson, *Proc. Roy. Soc.* (London) **A117**, 600 (1928).
77. L. de Broglie, *Recherches d'Un Demi-Siecle*, Albin Michel, Paris, 1976.

Index

absolute beginning, 39
action at a distance, 4
action function, 29, 32
affine connection, 7
algebraic logic, 10, 11, 15
analytic, 19, 26
analytic truth, 65–67
Andromeda, 2
Aquinas, T., 100
Aristotle, 99, 100
astronomy, 99, 101

beginning of the universe, 7
bending starlight, 13
Bentley III, R., 4
big bang, 6, 7, 10
black hole, 10, 60
blackbody radiation, 52
blueshift, 7
Bohr, N., 69
Born, M., 81, 116

cause-effect, 8
CERN supercollider, 25
closed system, 13, 84, 89–91, 104,
    109–112, 114, 118, 123
cluster of galaxies, 38
conservation of energy, 10
continuous field, 80–83, 85, 95–97,
    112–114
continuous group, 26
continuous set, 10

Copenhagen view, 89
Copernicus, 2
Cornu spiral, 48
correspondence principle, 117, 119,
    120, 123
cosmic radiation, 16
cosmological red shift, 37
cosmological time, 38, 49
cosmology, 1, 6, 7, 10, 11, 15, 17
covariance, 28
covariant derivative, 20
creationism, 67
current density, 34
curved spacetime, 7, 12, 13
cyclotron effect, 57

dark matter, 3
Davisson and Germer experiment, 78,
    115
Davisson, C.J., 78, 115
de Broglie, L., 78, 81, 115, 116, 123
delayed action-at-a distance, 58, 59
Democritus, 79, 95
Dirac equation, 118
Dirac spinor, 53
Dirac's electron equation, 15
discrete reflections, 26
discrete space time, 117
Doppler effect, 5

Eddington, A., 24
eigenvalue equation, 71

Einstein field equations, 11, 21, 23
Einstein group, 10, 13, 14
Einstein tensor, 18–22
Einstein's principle of relativity, 2
Einstein, A., 18–26, 95
electromagnetic field intensity, 34
electron diffraction, 79, 81, 115, 117, 123
elementary interaction, 70
empty space, 37
energy-mass relation, 13
energy-momentum tensor, 20
entanglement, 96
entropy, 68, 69
epistemology, 68, 69
equation of motion, 39, 40, 44
essential parameters, 10, 11, 14
Euclidean metric, 19
event horizon, 23
expanding universe, 5, 6
explanatory, 69

factorized field equations, 16
Faraday effect, 92
Faraday, M., 58
field equations, 9–11, 13, 14, 16, 17
fractional charge, 91
Fresnel integral, 48, 49

galaxy, 2, 5
Galileo, 1, 2, 8, 12
Galileo's principle of inertia, 12
Galileo's principle of relativity, 2
gauge transformation, 120
general relativity theory, 4, 7, 10, 11, 13, 17
generalized Mach principle, 71, 72
geocentric model, 99
geodesic, 12, 16
geodesic equation, 16, 21
geometric logic, 11
Germer, L.H., 78, 115
Gibbs, J.W., 68
gravitational red shift, 24
gravity, 18, 22, 24, 25

Hamilton, W.R., 16
Heisenberg, W., 69
heliocentric model, 101
Herschel, W., 2
Hesse, M.B., 4
Hilbert space, 89, 91, 112, 121, 124
holistic, 72
holistic medicine, 98
homogeneous matter, 38, 39
Hubble constant, 46
Hubble law, 5, 7, 11, 17, 18
Hubble telescope, 1

implosion, 39
inertial frame, 13
inertial mass, 89–91, 97, 106, 109–112, 115, 117–121, 123, 124
isotropic matter, 39

Janiak, A., 4

Kabbalah, 41
kaon, 88
Kepler, 3

Lagrangian, 29, 32
Lamb shift, 79, 87
language of cosmology, 17
language of mathematics, 66
laws of electromagnetism, 11
laws of motion, 2
laws of nature, 2, 4, 5, 8–10, 13, 14
lens-type telescope, 2
lepton, 53
Lie group, 10, 11, 14

Mach principle, 52, 70–73, 77, 90, 92, 97, 108–112, 115, 118, 121, 123
Mach, E., 52, 70–73
magnetic monopole, 34
Majorana equation, 117
mass, 80, 81, 88–91, 97, 102, 106, 109–112, 115, 117–124
mass doublets, 52
masses of elementary particles, 52

matter waves, 78, 79, 81, 123
matter/antimatter separation, 18
Maxwell's equations, 17
metric tensor, 11–13, 15, 19–21, 25
Milky Way, 2, 3

necessary truth, 8, 65
neutrino, 52
Newton, 2–4, 13
Newton's first law of motion, 40
Newton's second law of motion, 40
Newton's third law of motion, 111
Newtonian limit, 121
Noether's theorem, 10
Noether, E., 10, 26, 35, 55, 83, 114
non-commutative, 29
non-deterministic, 80
nonlinear equation, 22, 84, 105, 112, 114
nonlinear field theory, 71
normal science, 64

objectivity, 109
observed-observer, 70
Olber's paradox, 57–59
ontology, 68, 69
oscillating universe cosmology, 6, 7

pair annihilation and creation, 105
paradigm, 64, 65
Parmenides, 80, 95
particle physics, 1, 15
particle-antiparticle pair, 51–53, 56
Pauli matrices, 14
perihelion precession, 24
philosophical truth, 65
photon, 52, 58, 59
physics of the universe, 1, 15, 17
pion, 88
Planck, M., 58, 59
Plato, 66, 99, 100
Poincaré group, 10, 14
Popper, K.R., 8
positivism, 67, 71, 81, 93, 97, 117
Pound, R.V., 24
principle of covariance, 7

principle of equivalence, 40
probability calculus, 71
proper frequency, 58
pseudoscalar electromagnetic
    interaction, 87, 88
Ptolemy, 99, 100

quantum electrodynamics, 79, 87
quantum mechanical limit, 70
quantum mechanics, 15
quasar, 102
quaternion, 11, 14–17
quaternion field, 26
quaternion formulation, 18
quaternion metrical field equation, 32

radius of the universe, 45
Ram, M., 97
Raphael, 100
realism, 81, 93, 97
redshift, 7
reflecting telescope, 2
regular function, 12, 84, 115
regular polygon, 66
religious truth, 6, 66, 67
renormalization, 79
Ricci tensor, 19–22
Riemann curvature tensor, 19, 20
Riemannian geometry, 12
Rosen, N., 10
rotation of galaxies, 41

Sachs, M., 77, 97, 108
scalar curvature, 20, 21
School of Athens, 100
School of Pythagoras, 66
Schrödinger, E., 78, 79, 81, 86, 90, 96,
    118
Schwarzschild solution, 22, 24
science, 64–67, 69, 70
scientific revolution, 65
scientific truth, 6, 8, 64–67
second law of thermodynamics, 68
second rank spinor, 28, 30, 35

separation of matter and antimatter, 56, 57
single big bang, 38, 39, 49
singularities, 10
16-parameter Lie group, 28
space, 66, 73
spacetime, 5, 7, 9–14, 16
spacetime language, 10, 11
spacetime logic, 25
speed of light, 7, 9, 10
spin-affine connection, 30, 119
spin curvature, 31, 33, 34
spinor, 86–88, 90, 91, 119, 121
spinor variables, 15
Spinoza, 80, 97
spiral motion, 49
spiral universe, 103–106
stable solution, 23
statistical mechanics, 89
Stonehenge, 1
subjective knowledge, 69

tau meson, 53
tautology, 82
theory of inertia, 89–91, 112, 124

Thomson, G.P., 78, 115, 116
Thomson, J.J., 116
threads of truth, 64
time, 37–47, 49
top quark, 91
torsion, 16
total causation, 82, 112
truth, 64–67, 70

unified field, 27, 32, 35, 36
unified field theory, 11, 16, 17, 85, 86
universal gravitation, 2, 3, 13
universal interaction, 71
US Constitution, 67

vacuum equation, 21–25
variational principle, 16
virtual field, 88

wave mechanics, 79, 81, 86, 118
wave theory of light, 2
wave-particle dualism, 81, 117
weak interaction, 87
Weyl equation, 52